SSP
1/15/2022

MASS CUSTOMIZATION:
CHALLENGES AND SOLUTIONS

MASS CUSTOMIZATION:
CHALLENGES AND SOLUTIONS

Edited by

Thorsten Blecker
Hamburg University of Technology (TUHH)

&

Gerhard Friedrich
University Klagenfurt

 Springer

Thorsten Blecker
Hamburg University of Technology
Germany

Gerhard Friedrich
University of Klagenfurt
Austria

Library of Congress Control Number: 2006920573

ISBN-10: 0-387-32222-1 (HB)   ISBN-10: 0-387-32224-8 (e-book)
ISBN-13: 978-0387-32222-3 (HB)   ISBN-13: 978-0387-32224-7 (e-book)

Printed on acid-free paper.

Printed in the United States of America.

9 8 7 6 5 4 3 2 1

springer.com

TABLE OF CONTENTS

Platform Products Development and Supply Chain Configuration: an Integrated Perspective

Xin Yan Zhang and George Q. Huang

Modularity and Delayed Product Differentiation in Assemble-to-order Systems

Thorsten Blecker and Nizar Abdelkafi

A New Mixed-Model Assembly Line Planning Approach for an Efficient Variety Steering Integration

Stefan Bock

LIST OF FIGURES

LIST OF TABLES

PREFACE

Mass customization is one of the most important competitive strategies in the current economy. Its primary objective is to maximize customer satisfaction by providing tailor-made solutions with near mass production efficiency. Some researchers even postulated that mass customization is the single way to competitiveness in today's dynamic business environment. In effect, mass customization has proved to be very successful in certain business fields. Companies that succeeded to reap the benefits of the strategy by joining mass production and make-to-order capabilities could secure a competitive advantage and outpace their competitors. However, the failure of some implementation projects made it obvious that mass customization is still challenging and requires further research support from academia and industry. Especially, new scientific methods and tools are required to mitigate the difficulties encountered when putting the strategy into practice.

The contributions in this book represent leading edge papers and recent advances in mass customization. The results reported in the chapters not only contribute to the support of a successful implementation of mass customization but also stimulate further research and scientific work in the field. Throughout the chapters, the broadness and complexity of the field is shown. It becomes very obvious that mass customization challenges can only be mitigated if a joint effort of various disciplines such as business administration, computer science and engineering can be achieved.

The origins of this book lie in the International Mass Customization Meeting 2005 (IMCM'05) held by June 2005 at the University Klagenfurt, Austria where researchers from many scientific disciplines and practitioners acting in various industrial fields had the opportunity to present their researches and to discuss many issues related to mass customization. IMCM'05 has provided a platform for original research in the area and for the exchange of ideas and problem solving approaches across various fields. From the 39 blind-reviewed papers accepted for presentation at the conference, only 11 chapters have been selected for this book. The selection process of the best chapters was extremely difficult owing to the high quality of papers accepted for presentation. Furthermore, the authors of the selected papers were asked to revise and extend their papers so as to make the chapters accessible to a larger audience. Therefore, the book can be useful not only for researchers but also for practitioners and graduate students in operations management, management science, business administration and computer science. An additional chapter has been written in order to introduce mass customization by providing a literature review and discussing the state-of-the-art of the concept.

Our thanks go principally, of course, to the contributors as well as to the program committee of IMCM 2005 which is composed of Dipl.-Ing. Dr. Alexander Felfernig, Dr. Gerhard Fleischanderl, Prof. Dr.-Ing. Norbert Gronau, Dr. Albert Haag, Prof. Dr. Cornelius Herstatt, Prof. Dr. Patrick Horster, Prof. Dr. Bernd Kaluza, Dipl.-Ing. Dr. Eva-Maria Kern, Prof. Dr. Wolfgang Kersten, Dr. Frank T. Piller, Dipl.-Ing. Christian Russ, Dr. Martin Sonnenschein, and Prof. Dr. Franz Wotawa for their invaluable support during the review process and the selection of the best papers published in this volume. Very special thanks also go to Dipl.-Ing. Dipl.-Wirtsch.-Ing. Nizar Abdelkafi for his continuous support and help in formatting this book. In addition, we would like to acknowledge the efforts of Springer Publishing and especially the senior editor Gary Folven.

Thorsten Blecker
Full Professor in
Department of Business Logistics
and General Management
Hamburg University of Technology
(TUHH), Germany

Gerhard Friedrich
Full Professor in
Department of Computer Science in
Production
University Klagenfurt
Austria

Chapter 1

MASS CUSTOMIZATION: STATE-OF-THE-ART AND CHALLENGES

Thorsten Blecker and Nizar Abdelkafi
Hamburg University of Technology, Department of Business Logistics and General Management

Abstract: Mass customization refers to a business strategy that conciliates two different business practices, which are mass production and craft production. It aims to provide customers with individualized products at near mass production efficiency. In fact, there are many approaches in order to implement mass customization. In this context the customer order entry point plays an important role as an indicator for the degree of customer integration and customization level. Before embarking on a mass customization strategy, it is mandatory to examine if some critical success factors have already been satisfied. The most relevant factors are: customer demand for customized products, market turbulence, supply chain readiness and knowledge driven organization. However, the mere satisfaction of these factors need not necessarily lead mass customization to success. In effect, the entire process, which in turn consists of many sub-processes, has to be coordinated and managed in a suitable manner. The main sub-processes in mass customization are: the development sub-process, interaction sub-process, purchasing sub-process, production sub-process, logistics sub-process and information sub-process. However, state-of-the-art mass customization still has to face many challenges. We identify the external complexity and internal complexity as the main problems that may jeopardize the implementation of the strategy. External complexity can be referred to as the difficulties encountered by customers when they have to select adequate variants out of a large set of product alternatives. On the other hand, internal complexity is experienced inside operations and manufacturing-related tasks. It is toward solving these main problems that the researches reported in the different chapters of the book have been carried out. Finally, in the conclusions we provide an overview of each chapter, while pointing out its main contribution to research on mass customization.

Key words: Mass customization, state-of-the-art, mass customization processes, internal complexity, external complexity

1.   INTRODUCTION

The term mass customization (Pine, 1993; Tseng/Jiao, 2001) can be regarded as an oxymoron that joins two concepts seeming to be opposite at a first glance. Mass customization aims to conciliate two business practices, which are mass and craft (single-piece) production. Mass production manufactures low cost products by reaping the benefits of standardization and scale economies. On the other hand, craft production assumes a high level of individualization since the products are tailored to specific customer requirements. Researchers in strategic management have postulated for a long time the incompatibility of the manufacturing principles underlying both production concepts. The work of Porter (1998) on generic strategies is the best argument with regard to this issue. He postulates that companies simultaneously pursuing differentiation and cost leadership are stuck in the middle and cannot achieve strategic success. However, the emergence of mass customization has broken this common belief. Mass customization is a business strategy that aims to provide customers with individualized products and services at near mass production efficiency.

It is the incapability of mass production to prevail in some business fields, which has driven companies to seriously consider the alternative of customization at a mass scale. Mass customization enables companies to achieve an important competitive advantage by joining product differentiation and cost efficiency. In some business areas, customers are no longer looking for standardized goods and services but for products that are exactly corresponding to their requirements.

The concept of mass customization increasingly attracts the interest of scientists and practitioners. It is a fascinating approach whose benefits have widely been discussed in the technical literature. Some examples from the practice corroborate the applicability of mass customization, thereby demonstrating that the strategy is not just an oxymoron with no reference to reality. Other examples which comment on the failure of some mass customization applications prove that the strategy can turn into a nightmare if some important issues are not taken into account. In fact, there is a lack of research and common framework concerning the effective implementation of the strategy in the practice. For instance, the critical success factors of mass customization still represent a research topic that is not sufficiently explored by academia. In this chapter, we will deal with the state-of-the-art mass customization and main research areas relating to this topic. At the end of the paper, we notice the contributions of the different chapters of this book in filling the research gaps and bridging theory and practice.

2. MASS CUSTOMIZATION APPROACHES

Mass customization is not a pure but a hybrid manufacturing concept, which joins the efficiency of operations and differentiation by providing highly value added products. The starting point toward mass customization may be mass production or one-of-a kind manufacturing. Mass production companies work on the basis of a fully anticipatory model by making standard products to stock. Because of market pressures and customer demand for a broader product portfolio, mass producing companies may decide to shift to mass customization. On the other hand, one-of-a-kind manufacturers may choose to get into mass customization due to volume expansion and existing similarities between their end products. To examine the origin of mass customizing companies, Duray (2002) has carried out an empirical study, in which 126 companies from different industries participated. The study proves that the origin of mass customization companies is actually mass production or one-of-a-kind (custom) manufacturing. The research also shows that the financial performance strongly depends on the level to which the chosen mass customization approach matches the non-mass customized product line characteristics. In other words, firms starting out of a custom manufacturing are more successful if they involve their customers at very early stages when they pursue mass customization. Similarly, those with a mass production origin achieve a better financial performance if the integration of customers occurs at later stages in the production cycle.

In fact, firms may choose to provide their customers with different degrees of customization. With this regard, there are many attempts to provide classifications and taxonomies of mass customization practices. These classifications generally draw the level of customer involvement as a main criterion. Basically, the further upstream customers are involved in the production process, the higher the level of customization. For instance, by using the level of customer involvement in the value chain, Lampel/Mintzberg (1996) develop a continuum of strategies. According to the authors, the simplified value chain comprises four main stages, which are: design, fabrication, assembly, and distribution. As the level of customization increases, the point of the value chain at which the customer order enters is moved upstream. The authors define five strategies, which are: pure standardization, segmented standardization, customized standardization, tailored customization, and pure customization. The lowest level of customization (pure standardization) occurs if all stages of the value chain are standardized. On the other hand, firms achieve the highest degree of customization (pure customization) if customers are enabled to have an impact on the design process. The other strategies are intermediate forms, which are situated between the extreme levels. It is important to note that pure standardization and pure customization can be considered as mass customization strategies, only if certain conditions are fulfilled. Starting from the definition of mass customization, the entire process should involve both mass production and customization parts. A standard product that bears certain flexibility, so that it can be customized by the retail or customers themselves can be regarded as a mass customized product. For instance, car seats can be standardized. However, they may permit the driver to set the ideal seat position individually, so as to guarantee the maximum amount of comfort. In addition, providing a set of individual value added services around a standard product can also be regarded as a form of mass customization. On the other hand, pure customization can be seen as a form of mass customization, only if the process involves a mass production part, in which some components are standardized. For more classifications and taxonomies, readers can refer to the following references: Pine (1993), Pine/Gilmore (1999), Duray et al. (2000), Da Silveira et al. (2001), MacCarthy et al. (2003), and Piller (2003).

3.   SELECTED CRITICAL FACTORS TO LEAD MASS CUSTOMIZATION TO SUCCESS

The critical factors for leading mass customization to success refer to the necessary conditions that if satisfied, the implementation of the strategy has great chances to be beneficial for the company. Before shifting to mass customization, it is important to examine if these factors are fulfilled. This discussion is driven by the fact that mass customization may not be suitable for each business environment or that the environment is not yet ready to facilitate the application of the strategy.

3.1 Customer Demand for Customized Products

The main driver for the implementation of mass customization is the customer. It is obvious that if the customer does not look for individualized goods and services, the strategy will not achieve success. Pine/Gilmore (2000) note that customer satisfaction and "voice of the customer" surveys enable companies to understand the general needs of the customer's base. But they are not adequate to help companies make decisions concerning if they should mass customize or not. In this context, in order to evaluate if the customer actually needs customized products, Hart (1995) uses the concept of "customer customization sensitivity", which is based on two basic factors, namely: the uniqueness of customers' needs and level of customer sacrifice. The uniqueness of customers' needs to a great extent depends on the type of the product in question. In effect, for some products, the customer may be indifferent to the broadness of the offered variety, and thus customization does not really make sense. On the other hand, the value of some other products is tightly related to the level of individualization that is provided to the customer. For instance, it is obvious that a hair cut or investment counseling are products that should be tailored to the customer's specific needs. The second factor: customer sacrifice is defined by Pine/Gilmore (2000, p. 19) as "the gap between what a customer settles for and what he wants exactly." In this context, there is a difference between what the customer would accept and what he really needs. As the extents of customer sacrifice and needs' uniqueness increase, the customization sensitivity increases; in other words, customers are readier to accept customized products. In the attempt to develop a standard framework for the assessment of customers' susceptibility to preferring customized products, Guilabert/Donthu (2003) developed a scale consisting of six main dimensions. These dimensions represent statements which are derived and tested empirically. For instance, the second dimension determined by the authors is as follows: "I wish there were more products/services that could be easily customized to my taste."

3.2 Market Turbulence

In addition to customer customization sensitivity, the market turbulence (Pine, 1993) is an important factor that the firm has to consider in order to determine the adequate point in time for the shift to mass customization. "The greater the market turbulence, the more likely that the industry is moving toward mass customization, and that the firm *has* to move in order to remain competitive" (Pine, 1993, pp. 54-55). It is important to note that the reasons of the success of mass production are mainly due to stable business environments, in which customers do not demand for differentiation. As market turbulence increases, these conditions are no longer satisfied, thereby triggering the failure of mass production to cope with the environmental changes. While Pine (1993) postulated that mass customization will completely replace mass production, Kotha (1995, 1996) demonstrates by means of a case study that both strategies can be implemented successfully by the same company. There are even synergy effects that can arise if both systems can interact properly.

3.3 Supply Chain Readiness

Nowadays, it is well argued that competition takes place between supply chains rather than between single companies (Christopher, 2005). In effect, firms increasingly reduce the level of vertical integration by focusing on their core competencies. Therefore, the role of suppliers becomes more important than ever. The shift to mass customization not only affects the internal operations but also the relationships between the company and its partners in the supply chain. For instance, in the event that strategic suppliers do not have the necessary capabilities and skills to support the mass customization process, the strategy will not be able to achieve its objectives. Furthermore, mass customization calls for a dynamic network, which should be configured according to customers' requirements in order to ensure high responsiveness. Thus, the alignment of the entire supply chain with the strategic objectives of mass customization is of high relevance.

Basically, the company serving the end customer bears the largest responsibility in leading mass customization to success or failure. In effect, it is the member in the supply chain that initiates such a strategic move. Furthermore, it is affected the most by the change since it has to devote large efforts to adapting and redesigning internal operations. Two main internal capabilities are necessary in order to customize products at low costs. The first is to design products in such a manner that they can easily be customized. In this context, the product architecture to a great extent influences the customizability degree. The second capability refers to process flexibility (Zipkin, 2001), which is required to manufacture individualized products on a mass scale. Especially, the manufacturing system should enable quick and smooth changeovers between products in order to minimize setup times and costs. Ahlström/Westbrook (1999, p. 263) point out that the offer of customized products for the mass market "…means not merely making operational adjustments for specific orders but developing process which can supply very numerous customer-chosen variations on every order with little lead time or cost penalty."

3.4   Knowledge Driven Organization

To accommodate mass customization, the organization should create an atmosphere, in which knowledge can be shared smoothly. Since the strategy aims to fulfill individual requirements, the input of customers should be managed effectively and translated into products and services. Thus, before shifting to mass customization, companies have to ascertain if they have the required capabilities ensuring that customer knowledge adequately flows in the organization. In addition, as described by Pine et al. (1993), processes in mass customization should be organized in loosely coupled modules. The only way to coordinate this highly flexible structure is an efficient communication and exchange of knowledge. Another important point is that mass customized products are high value added products, which are intrinsically knowledge intensive.

4.   THE MASS CUSTOMIZATION PROCESS

The mass customization process can be defined as the set of interlinked activities that are necessary to capture individual requirements, to translate them into the physical product, which is then produced and delivered to the customer. The customization process can be divided into many sub-processes including the main stages of the value chain. Blecker et al. (2005) has identified six sub-processes, which are: the development sub-process, interaction sub-process, purchasing sub-process, production sub-process, logistics sub-process, and information sub-process.

4.1   Development Sub-Process

As mentioned earlier, the customizability of the product is an important issue in order to lead mass customization to success. If the product is not developed in such a way that it can easily be adapted to the customer requirements at low costs, mass customization would never be able to achieve its goals. Thus the role of the development sub-process is to translate different needs of diverse customers into generic product architecture from which a large number of product variants can be derived. In this way, each customer specific product can be considered as a particular instantiation of the generic design.

A product architecture is "…the scheme by which the function of the product is allocated to physical components" (Ulrich, 2003, p. 118). It is widely argued that modular architectures to a great extent facilitate the customization of the product (e.g. Pine, 1993). Modularity ideally involves a one-to-one mapping from the elements of the function structure to product building blocks. In addition, the interfaces between modules are well-defined and de-coupled. The creation of product variety occurs by mixing and matching the modules into different configurations. The benefits and limitations of product modularity have already been discussed by many authors in the literature (e.g. Pine, 1993; Piller, 1998; Baldwin/Clark, 2000; Garud/Kumaraswamy, 2003; Langlois/Robertson, 2003). Modularity enables companies to achieve the economies of scale, economies of scope, and economies of substitution. Furthermore, it considerably reduces development lead times. Some disadvantages of modularity may be the ease of imitation by competitors and costly development.

Besides the modularity of products, other concepts such as commonality and platform strategies are of high relevance in order to increase reusability in mass customization. Commonality (e.g. Collier, 1982; Baker, 1985; Wacker/Treleven, 1986; Eynan/Rosenblatt, 1996) refers to the multiple use of components within the same product and between different products. It aims to reduce the extent of special-purpose components which generally increase internal variety and costs. Design engineers usually find it easier to design a new component than to search in large databases for available components that can satisfy the particular design problem. Therefore, companies have to put policies so as to encourage design teams to use existing components in new product development.

The combination of commonality and modularity leads to the product platform strategy (Nilles, 2002). We refer to a product platform as a common module that can be implemented into a wide range of end variants of the product family. Meyer/Lehnerd (1997, p. xiii) define a product platform in the broader sense as "...*a set of subsystems and interfaces that form a common structure from which a stream of derivative products can be efficiently developed and produced.*" Platforms are generally cost-intensive and should be well-planned in order to gear up the development of future derivative products.

Thus in mass customization, product development is mainly carried out by design engineers who define the degrees of freedom in product design (i.e. a set of module options) that customers can exploit in order to create individualized product variants. However, a higher level of customer integration can be achieved if the product development function itself is moved to the customer who becomes much more involved in product innovation. In other words, customers will be given the possibility to develop new product concepts by themselves or co-develop it with the innovating company. Toward this aim, companies can provide their customers with toolkits for open innovation (von Hippel, 2001), which are software systems that help users overcome their limitations in realizing product ideas and visions. For instance, Piller et al. (2004) report about a toolkit, which makes it possible for customers without any software programming skills to create mobile phone games on a desktop computer. Therefore, toolkits can be considered as adequate means to exploit the latent customer energy (Sonnenschein/Weiss, 2005) and customer's innovative abilities.

4.2 Interaction Sub-Process

The output of the development sub-process is the so-called solution space which consists of the set of product alternatives that can be produced by the mass customizer. In the attempt to increase the chance that each customer finds the product that exactly fulfills his requirements, the firm tends to develop a very extensive solution space with sometimes billions of product variants. Therefore, there is a need for a sub-process that matches the customer's expectations with physical products. In fact, this is done during the interaction sub-process which captures and identifies what the customers need and assigns to their requirements the most appropriate end product. Zipkin (2001) refers to the interaction process as the elicitation process. He points out that there are four kinds of elicited information in mass customization, which are: identification (e.g. name and address); customers' selections from menus of alternatives; physical measurements; and reactions to prototypes.

Basically, the interaction process between the customer and company can be supported by the retail or carried out directly over the Internet. For instance, to be able to sew an individualized garment in the clothing industry, the body measurements of the customer are required. The acquisition of such data may occur through optical body scanners that are located at the retail stores. In this case the retail plays an important role in the customization process since it captures the customer's requirements and communicates them to the producer. However, in order to buy a customized personal computer, customers may not have to go the store. They can log onto the website of a computer manufacturer such as Dell and then configure the product according to their needs. The interaction process completely runs via the electronic medium.

In mass customization, customers generally do not have a passive role as in mass production but they actively participate in the value adding process. Because of this input into the value chain, customers are often called "co-producers" or "prosumers" (Toffler, 1980; Piller, 2003). The extent of customer integration may vary from the simple selection out of predefined alternatives (in the event that the mass customizer pursues a customized standardization) to the co-design of products (when the mass customizer follows a pure customization strategy).

4.3 Purchasing Sub-Process

Due to the decreasing level of vertical integration, suppliers can be considered as a relevant source of competitive advantage and a potential leverage for cost reduction. The coordination of the outsourcing process is allocated to the purchasing team which negotiates with suppliers, selects the best ones and may place long term contracts with them. Therefore, the purchasing department is of high importance in the value chain since it represents the interface which links the company with its upstream suppliers. Managers and researchers argue that the achievement of high profits lies in an effective and efficient component and material purchasing. Successful assemble-to-order companies such as Dell rely on advanced network of suppliers which deliver components that represent a large percentage of the total value added of the product. In a mass customization environment, the purchasing department should ascertain that suppliers have the required responsiveness and flexibility in providing variety.

The design of products around modular architectures has led companies to rethink their outsourcing strategies. Modular sourcing has emerged to reconcile the conflicting objectives of decreasing the level of vertical integration and reducing the suppliers' base. While the traditional view consists in delivering discrete components to the manufacturer, modular sourcing aims to outsource entire complex modules to the suppliers. The modular sourcing concept enables the mass customizer to reduce the complexity of the purchasing process. However, the approach calls for trust in the supply chain and a close partnership with module suppliers. According to Sako/Murray (1999) there are basically two approaches to apply modular sourcing. The mass customizer can be the integrator if it retains the module control or it can be the modularizer if the module control is completely transferred to the supplier that possesses the capabilities required to provide modular solutions. Doran (2003) investigates the implications of modularization on the supply chain in the automotive industry. He proposes a classification of first-tier suppliers that involves mature, developing, and fringe first-tier suppliers. Whereas mature and developing first-tier suppliers possess the capabilities required to produce complex modules, fringe first tier-suppliers did not develop the necessary skills and competencies and are likely to move down in the structure of the supply chain to become second-tier suppliers. From this, it follows that the purchasing function in mass customization has to assess carefully the capabilities of suppliers if they can develop and assemble complex modular solutions since not all suppliers are abele to achieve this task.

4.4   Production Sub-Process

Pine (1993) illustrates the importance of manufacturing flexibility for mass customization by showing that the economic order quantity (EOQ) takes up low values (in the extreme case equal to one) if the set up costs are considerably reduced (ideally to zero). Therefore, in order for mass customizing companies to produce variety in an efficient manner, changeover activities that are necessary to change parts, fixtures, tooling, equipment programming from one product to another (Anderson, 1997) must be minimized. In the technical literature, the feasibility of mass customization in practice is mainly ascribed to the advances realized in the fields of flexible manufacturing systems and modular product architectures (Piller/Ihl, 2002). In this context, Duray et al. (2000) even postulate that product modularity is a main building block of manufacturing environments traditionally regarded to be flexible.

However, it is important to note that modularization may not be a necessary condition for customization. In effect, examples from the practice document on the possibility to customize products such as garments or shoes, which are intrinsically integral. For instance, the clothing industry uses computer-controlled machines for cutting fabrics according to individual body measurements. Therefore, two types of production systems for mass customization can be distinguished according to the source of flexibility. The first type relies on flexibility that is built into the product design through modularity, whereas the second type depends on flexibility that is built into the process. With this regard, it can be stated that product modularity facilitates the customized standardization approach where standard modules are assembled on the basis of customer requirements and process flexibility makes it possible to pursue a tailored customization strategy, in which the customer order enters at the fabrication stage.

The entry point of the customer's order in the production process determines the customization level that is offered by the firm. This point is typically called the decoupling point or differentiation point. It is generally defined as the point in the production process at which the products assume their unique identities. Furthermore, it is situated at the boundary between the push and pull systems. In fact, the determination of the optimal placement of the decoupling point is not an easy decision. Many considerations should be taken into account. The mass customizer should ascertain the product variety that is required by customers and more importantly the costs incurred by this variety. In this context, delayed product differentiation and postponement are two important key terms. They are two related concepts, whereby the first means placing the decoupling point at later stages in the production process and the second describes that some production activities are not initiated until customer order arrives.

4.5   Logistics Sub-Process

This process involves the upstream logistics with suppliers and downstream logistics with customers. The upstream logistics deals with the transportation, consolidation and warehousing of materials and components that are required for production. On the other hand, downstream logistics ensure the packaging and shipment of end products to customers. Both upstream and downstream logistics face enormous challenges in mass customization. The upstream logistics should ensure that components and modules are delivered in time according to the mass customizer's schedule. The downstream logistics has to deliver on a per item basis when customized products are directly shipped to the customer. This considerably increases logistics costs since each customer is served individually. Furthermore, poor delivery reliability would make customers doubt about the benefits of mass customization. Riemer/Totz (2001) point out that the downstream logistics may carry out a part of the customization process if the customer is provided with the possibility to choose form different logistics options of packaging and transport. Customized packaging (e.g. gift wrapping) or individual delivery times are just few examples, which illustrate the involvement of logistics in the process of customization.

Logistics generally calls for large investments in transportation and warehousing equipment. Especially, for mass customization the distribution of customized products can incur high costs, which considerably increases total product costs. Because of this, there is a growing tendency of companies towards outsourcing their logistics operations to third-party logistics (3 PL). Third party logistics are suppliers of logistics services that create value for their commercial clients. They have elaborated transportation networks and can achieve the economies of scale in logistics by consolidating orders from different industrial customers. The services offered go beyond transportation and storage to include value added services, e.g. customized packaging or even final assembly of products (e.g. Lee 2004, van Hoek 2000). Some authors, e.g. Gunasekaran/Ngai (2005) mention the potential role of fourth party logistics (4PL)[1] providers in the context of product customization. 4PL refers to the integration of all companies involved along the supply chain. In fact, there is an endeavor of many logistic companies such as DHL, FedEx, and UPS to provide such services by linking and coordinating the members of the supply chain on the basis of their information and communication systems.

Most often, mass customization (except for pure standardization in the classification framework of Lampel/Mintzberg (1996)) does not trigger inventories at the end product level. Common components have steady demand due to the risk pooling effect, while less frequently demanded (special-purpose components) have an uneven demand. Chandra/Grabis (2004) mention that such components require different approaches to inventory management. The authors suggest using MRP-based policies for the management of components with variable demand and reorder-based policies for managing globally sourced components with variable demand. However, JIT policies are suitable for the management of locally sourced components with even demand. In fact, inventories to a great extent indicate the degree to which logistics function smoothly. "Most things that go wrong in a logistics system cause inventory to increase" (Tesrine/Wacker, 2000, p. 114).

---

[1] The term "4PL" was actually coined by the consulting group Accenture, which also holds the trademark to this name.

4.6 Information Sub-Process

The information sub-process interacts with all processes mentioned earlier. It aims to integrate the main activities required to customize products by ensuring a smooth information flow (Blecker/Friedrich, 2006). An effective integrated information system for mass customization should capture the customer needs, develop a list of product requirements, determine manufacturing specifications with respect to routing, material processing, assembly, etc. Furthermore, it should offer the possibility to set up the manufacturing system, arrange for end product shipment, and enable the verification of the product's order status (Berman, 2002).

In order for companies to practice an efficient customization, products have to be identified at the single product level. In this way, items can be controlled individually along the supply chain, e.g. in manufacturing and distribution (Kärkkäinen/Holmström, 2002). A promising technology that makes such identification possible is radio frequency identification (RFID), which not only enables the storage of product specific information but also real-time modification of this data during product processing. For instance, RFID is regarded as an important enabler of QSC Audio Products to move from make-to-stock to mass customization. With the aid of this technology, the assembly line receives the information required to control assembly work and routings from an RFID tag attached to the product itself (Feare, 2000). Although the benefits of RFID technology in streamlining information flow in mass customization supply chains are obvious, the costs of the technology are still high, thereby making companies reluctant to adopt it.

Another useful approach for the coordination throughout the mass customization supply chain is vendor managed inventory (VMI). This concept is enabled by electronic data interchange (EDI) and considerably facilitates inventory replenishment because the supplier can retrieve real-time data about stock levels of modules and components. In addition, the integration of ERP (Enterprise Resource Planning) systems among the main members of the supply chain improves the agility and adaptability to unforeseen events. In this way, if unexpected changes arise, the suppliers can immediately react and adjust their activities. Mass customization would also profit from the advances in software engineering, e.g. service-oriented architectures (SOA) with the objective to couple information systems of different partners in a loose manner through the use of standardized interfaces and services. Mass customization also calls for internal information systems such as product data management systems which should be very sophisticated in order to cope with the extensive variety induced in these environments.

5.   MASS CUSTOMIZATION CHALLENGES

There is no doubt concerning the benefit of mass customization as a strategic concept, which enables companies to outpace competitors. However, merely recognizing the benefit need not necessarily mean a successful implementation of the strategy. Many customers are still reluctant to buy customized products and companies are also skeptical about the feasibility of the strategy in practice. The difficulties in implementing a successful mass customization are mainly due to two main problems, which we call them external and internal complexity. External complexity refers to the uncertainty encountered by customers when they intend to customize their products. On the other hand, internal complexity is experienced inside the company's operations. It refers to the problems faced by the company because of the extensive product variety induced in mass customization. In the following, we will deeply analyze these problems and explain how they can jeopardize the success of mass customization.

5.1   External Complexity

Customers generally do not look for choice per se; they do only want the product alternatives that exactly fulfill their requirements. In large variety environments, customers generally feel frustrated, confused and incapable of meeting optimal decisions. They are overwhelmed by the product selection process and experience so-called external complexity. This external complexity arises because of three main factors, which are: (1) the limited information processing capacity of humans, (2) lack of customer knowledge about the product, and (3) customer ignorance about his or her real individual needs. The first complexity driver is inherent to human beings and cannot be influenced. In effect, the ability to perceive a large number of options and to compare between them requires processing capabilities that humans do not possess. For instance, Miller (1956) has ascertained that the capacity of humans to simultaneously receive, process, and remember data is limited to seven units (plus or minus two). The second complexity driver arises if customers lack knowledge and expertise about the product. As customers become more accustomed with the product; their capabilities of making rational comparisons between options get better. Customers with good product knowledge can grasp the product functionalities and reduce the solution space to a manageable subset from which they make optimal choice. For instance, customers who already have used PCs or mobile phones would find it easier to choose the most suitable PC or mobile phone than those who never have used these products. The third complexity driver refers to the difficulty encountered by customers when they intend to estimate and describe their needs. "Customers often have trouble deciding what they want and then communicating or acting on them. [...] There are situations in which customers clearly articulate their requirements. More commonly, however, customers are unsure" (Zipkin, 2001, p. 82). In this context, Blecker et al. (2005) develop a model which demonstrates that customers in a mass customization environment actually order products that are not fitting their requirements. This model distinguishes between two types of needs, which are: the subjective and objective customers' needs. The subjective needs are defined as the individually realized and articulated requirements whereas the objective needs refer to what customers really want. On the basis this model, the authors suggest that customers should be provided with a good support during the interaction process in order to help them identify their objective needs

In the attempt to facilitate the product search task, companies provide their customers with online software tools called configuration systems. Given a set of customer requirements and a logical description of the product family, the role of the configuration system is to find valid and completely specified product instance along all of the alternatives that the generic structure describes (Sabin/Weigel, 1998). A configuration system can also automate the order acquisition process by capturing customer requirements and transmitting them to production without involving intermediaries. The advantages of configuration systems are discussed by many researchers in the literature (e.g. see Forza/Salvador, 2002). However, these benefits seem to be more obvious in the Business-to-Business field. In the Business-to-Consumer field, configuration systems still need substantial improvements. Especially the way by which configuration systems present product options to customers should be more emphasized (Blecker et al., 2005). In effect, the satisfaction of customers with their final choices to a great extent depends on the manner by which product information is presented (Huffman/Kahn, 1998). If customers are overwhelmed by the configuration task, they may abort the configuration process and logout of the website or make suboptimal choice decisions (Piller, 2003). An empirical study carried out by (Rogoll/Piller, 2002) demonstrates that there is no configuration system that fulfills optimal requirements from company's and customer's perspectives. More importantly, configuration systems are not able to solve the external complexity problem (see e.g. Blecker et al., 2004). Von Hippel (2001) mentions that current configuration systems just enable customers to select products out of alternatives but do not facilitate customer learning. Customer learning is to provide customers with suitable tools in order to ascertain before placing customized orders that the product actually corresponds to their real requirements. The reduction of external complexity is a challenging issue in mass customization. If the customers do not order the right products, their trust in mass customized solutions will decrease drastically. Well-designed configuration systems, however, enable companies to reduce the level of external complexity so as to implement mass customization more effectively.

5.2 Internal Complexity

In addition to the external complexity problem, mass customization induces so-called internal complexity. Internal complexity is mainly due to the proliferation of product variety which negatively affects operations by increasing costs and slowing down the velocity of the supply chain. Anderson (1997, 2004) makes the distinction between external and internal variety. External variety is seen by customers, whereas internal variety is experienced inside manufacturing and distribution operations. The relationship between external and internal variety can be illustrated by means of Ashby's law of requisite variety. This law states that "variety can destroy variety" (Ashby, 1957, p. 207). From an operations' perspective, this means that the fulfillment system should have sufficient internal variety (processes, components, fixtures, tools, etc.) in order to represent and control the external variety required by customers. It is not possible to serve customers with large variety without keeping a certain variety inside the system.

An extensive product variety in mass customization cannot be manufactured without a certain loss of efficiency. For instance, an empirical study of Wildemann (1995) has shown that with the doubling of the number of products in the production program, the unit costs would increase about 20-35% for firms with traditional manufacturing systems (Job shop systems). For segmented and flexible automated plants, the unit costs would increase about 10-15%. Wildemann concluded that an increase of product variety is associated with an inverted learning curve.

Yeh/Chu (1991) developed a theoretical framework in order to examine the effects of variety on performance. In their framework, they distinguish between the direct and indirect effects of variety. The direct effects include product flexibility, part and process variety, number of set ups, inventory volumes, material handling, and production scheduling. However, the indirect effects mainly relate to delivery reliability, quality and costs. The empirical test of the framework shows that product proliferation has a significant impact on cost and a moderate impact on service and delivery times. The significant impact of variety on the costs' position of companies is due to manufacturing complexity which mainly arises in the form of overheads.

Rosenberg (1996) illustrates the rapid proliferation of product variety by means of a simple example from the automotive industry. He shows that the number of combinations of modules into car variants can reach astronomical scales (billions of alternatives) by only using a few must-modules (modules that are required to fulfill the basic functionalities of cars) and relatively small number of can-modules (optional modules). It is worth noting that the example provided by Rosenberg is still conservative because variety proliferation as actually experienced in the practice can be even larger. Though modularity lowers product complexity and reduces the number of parts and subassemblies, the challenge in such mass customization environments mainly relates to the planning and scheduling of production. With increasing product variety, the complexity of the scheduling function increases because of additional product changeovers, more routing alternatives on the shop floor, larger volumes of work-in-process inventories, assembly line balancing problems, increasing variability, etc. In this context, Byrne/Chutima (1997) point out that the use of flexible production systems does not solve the entirety of the problem; it even may aggravate it. In effect, each added degree of freedom due to production flexibility increases the size and complexity of the scheduling function.

With the proliferation of part variety, the complexity of purchasing increases. In purchasing, supplied parts and materials are not all handled the same way. Companies generally use specific criteria to classify purchased parts (e.g. generics, commodities, distinctives, criticals (Coyle et al., 2003)) and accordingly determine the type of the purchasing process and nature of relationships with suppliers. Hence as variety increases, the count as well as the type of purchasing processes to be planned, managed and controlled increases. In addition, the suppliers' base may get bigger, which would make the coordination of the entire supply chain more difficult. With this regard, suppliers' selection is a major issue because it not only determines the mode of collaboration in the future but also the structure of supply chain configuration and thus the network complexity. Although Just-in-time policies can alleviate the negative effects of the network complexity, it can be implemented only with local suppliers to manage the delivery of components with steady demand. If these conditions are not satisfied, inventory policies which involve safety stocks are used. Since variety negatively impacts the stability of components' demand, Just-In-Time policies are no longer applicable, thereby increasing the complexity of the network and thus the stock levels in the supply chain. Furthermore, variety in mass customization increases complexity not only at the supply side but also at the distribution side. Complex distribution networks are required to deliver individual products on a per item basis and more importantly to provide an effective and efficient an after sales service. The effects of variety-induced complexity on the performance and productivity of the firm are depicted in many contributions in the literature (see e.g. Perkins, 1994; MacDuffie et al., 1996; Fisher/Ittner, 1999; Ramdas, 2003).

6. CONCLUSIONS AND CONTRIBUTION OF THE BOOK

In this chapter, we have depicted the main approaches to implement mass customization. These approaches are generally classified according to the degree of customer involvement in the value chain. After that, we dealt with the critical factors that should be satisfied before a company shifts to mass customization. The discussion makes it obvious that managers should consider many important issues when they want to embark on the strategy. However, the fulfillment of these critical factors is a necessary but not a sufficient condition in order to lead mass customization to success. To deliver high value added products to the customer on mass scale, the mass customization process must run adequately. This calls for a smooth interplay between a set of sub-processes (development, interaction, purchasing, production, logistics, and information). We kept these sub-processes as generic as possible, so that they can fit many mass customization environments. The implementation of the strategy in the practice still faces many challenges in the practice. These are mainly due to the external and internal complexity problems. External complexity refers to the uncertainty and confusion encountered by customers during the product selection process. Internal complexity is experienced inside operations because of the variety of products, parts, and processes that are induced by mass customization. The following chapters of this book address both challenges and describe several solution approaches in order to make mass customization work efficiently.

In the following chapter, *Kreutler and Jannach* address the external complexity challenge in mass customization. The authors argue that the typical "one style-fits-all" approach for needs elicitation over the Internet is not adequate for customer-supplier interaction in a mass customization environment. Instead, the elicitation process should be adapted to the customer not only at the content level but also at the interaction and presentation level. To achieve this goal, adequate techniques for an effective implementation of the necessary functionalities in online configuration are discussed. Furthermore, the required architectural aspects to build and maintain such highly adaptive applications are provided. Thus, the authors extend state-of-the-art configuration systems which lack adaptability and customer-oriented support.

Configuration systems have been widely used to support mass customization environments where the products are physical. In fact, when we think about customized goods, computers, cars, shoes or garments come usually first to mind. However, as it will be shown by *Wolter, Hotz, and Krebs*, the mass customization concept proves to be applicable for intangible products such as software or software-intensive systems consisting of both hardware and software. The authors develop a tool that enables an automatic derivation of the software artifacts required to fulfill certain features specified by the customer. The presented approach even supports the evolution of software product families in the course of time. In effect, by means of a dependency analysis, a method is provided to compute the impacts of the modification or addition of certain software components on others. A prototype implementation with industrial partners provides interesting and promising results concerning applicability in practice.

Configuration systems either for physical or for software products require a product model, which can be regarded as the core that contains information about the different (physical/software) components to be assembled as well as the constraints existing between them. Well designed product models avoid redundancies so that no data is stored more than once. Toward this aim, *Jørgensen* develops a model with multiple abstraction levels for the representation of product families. By means of this new approach, the author shows that product configuration can be shifted toward identification and definition of attributes instead of modules and components. In this context, the functions that are required by customers are addressed at higher abstraction levels.

Enabling the customer to find the right product alternative that meets his or her requirements does not mean by far that mass customization can be implemented efficiently. In effect, mass customization calls for an optimal balancing between customer preferences and operational reality. The achievement of operations efficiency is highly dependent on product design and the extent of induced internal variety. In the attempt to reduce internal variety, *Ismail, Reid, Poolton and Arokiam* develop a methodology based on so-called similarity matrix and similarity coefficients. The implementation in the practice shows that the approach can be used for the redesign of existing product families, so as to increase feature reusability. The similarity coefficients also help design engineers make rational decisions concerning the design features to be used when developing new products.

However, mass customization not only calls for the redesign of products to make them customizable on a mass scale. There are also process requirements that should be satisfied by the production system, so as to be able to manufacture a large product mix while reducing setup activities on the shop floor. In this context, *Reik, McIntosh, Owen, Mileham and Culley* propose a systematic method called "design for changeovers" that aims to analyze manufacturing equipment with respect to setup activities. The method enables the initiation of manufacturing process improvements whose positive impact on changeovers can be evaluated by means of a set of metrics.

In the attempt to analyze the mutual impacts of platform products development and supply chain configuration, *Zhang and Huang* propose a comprehensive research framework based on three main dimensions. The first dimension refers to the integration levels or the coordination schemes of supply chain agents. The second dimension consists of the product platform development strategies such as commonality, modularity, postponement and scalability. The third dimension is about supply chain configuration decisions, which include e.g. supplier selection, inventory allocation, ordering policy, operation selection, service time, etc. Then, the authors investigate in detail a specific case, in which the agent relationship is non-interactive, commonality and modularity strategies are considered, and supply chain configuration decisions include supplier selection, inventory allocation, and ordering policy. The authors formulate a non-linear mathematical decision model for this problem and present a solution procedure. On the basis of a numerical example, it is shown that platform products strategies decrease total costs and lead manufacturers to choose module suppliers with a high level of specialization.

In addition to their impacts on the supply chain configuration, product architectures necessarily affect the internal manufacturing system. In this context, *Blecker and Abdelkafi* examine the relationships between the modularity of products and delayed product differentiation on the shop floor. The main thesis of their work is that the principle of delaying product differentiation turns to be insufficient if products are modular. In other words, the mere application of this principle does not enable one to ascertain in a suitable way how variety should increase within the assembly process. In order to determine the optimal proliferation of variety, the authors propose to additionally consider the variety-induced complexity. The weighted Shannon entropy is suggested as a measure for the evaluation of this complexity. Furthermore, the authors explore the complexity measure and its managerial implications in the case of a two-stage assembly process.

Whereas *Blecker and Abdelkafi* concentrate on the optimal proliferation of variety in the assembly process, *Bock* focuses on assembly line balancing in the presence of large product variety, He develops a new balancing approach based on a modular variant definition. The resulting model can deal with theoretical variant programs comprising several billions of end product variants. In addition to this, the author develops a randomized parallel Tabu search algorithm in order to determine appropriate line layouts in a systematic way.

Brabazon and MacCarthy take a broader view which goes beyond the assembly process on the shop floor to involve the entire fulfillment process. The authors review the literature dealing with the order fulfillment models for the catalog mode of mass customization. Catalog mass customization is defined as a concept in which customer orders are fulfilled from a pre-engineered set of potential product variants that can be produced with a fixed order fulfillment process. The literature review reveals that four main order fulfillment structures have been developed thus far. The authors distinguish between fulfillment from stock, fulfillment from a single fixed decoupling point, fulfillment from one of several fixed decoupling points and fulfillment from several points, with floating decoupling points. Furthermore, they discuss the main influences on the choice of fulfillment mechanism, namely: product variety, postponement, process flexibility, fulfillment logic, and customer factors. Thus, the contribution of the authors is a comprehensive framework, which consists of the possible concepts that can be used in order to put catalog mass customization into practice.

Optimal order fulfillment in mass customization to a great extent depends on inventories and corresponding stock levels. In this context, *Lu, Efstathiou and del Valle Lehne* develop a lean inventory model for mass customization. By using combinatorial mathematics, the authors express the customer service level as a function of product variety and inventory capacity. In addition, they show how to adjust product variety and inventory capacity in order to meet a targeted customer service level. The developed function enables managers to ascertain how much inventory capacity should be extended in the event that the number of product variants increases and service level should be kept constant. Similarly, given a certain number of product variants, it is possible to derive the required increase in inventory capacity in order to improve the service level.

The book chapters that are mentioned until now deal with issues that are related to organization and technology in the specific context of mass customization. A fundamental topic, which is as important as organization and technology for the industrial firm in general and mass customizing enterprise in particular, concerns the human resources. In spite of its high relevance, the issue of personnel in mass customization has been neglected in research. Therefore, in the attempt to fill this research gap, *Forza and Salvador* have conducted a qualitative study based on experts' interviews with the objective to explore the individual competences supporting the organization capability for mass customization. The research framework developed by the authors assumes that mass customization capability depends on competence, which is in turn a construct that is made of three classes of individual characteristics, namely abilities, knowledge and attitudes. Within this research framework, the authors explore what roles are mostly affected by mass customization within the company and what fundamental requirements mass customization poses on the manufacturing firm in terms of individual competences.

Recapitulating, the contributions provided in this book handle many issues that are necessary for making mass customization work efficiently. Throughout the chapters, the authors provide theoretical concepts, tools, and practical methods in order to streamline the mass customization process, while mitigating the negative effects triggered by the internal and external complexity.

REFERENCES

Ahlström, Pär / Westbrook, Roy (1999): Implications of mass customization for operations management: An exploratory survey, International Journal of Operations & Production Management, Vol. 19, No. 3, pp. 262-274.

Anderson, David (1997): Agile Product Development For Mass Customization, Chicago-London-Singapore: IRWIN Professional Publishing 1997.

Anderson, David (2004): Build-to-order & Mass Customization – The Ultimate Supply Chain Management and Lean Manufacturing Strategy for Low-Cost On-Demand Production without Forecasts or Inventory, Cambria, California: CIM Press 2004.

Ashby, Ross, W. (1957): An Introduction to cybernetics, 2nd Edition, London: Chapman & Hall LTD 1957.

Baker, Kenneth R. (1985): Safety stocks and Component Commonality, Journal of Operations Management, Vol. 6, No. 1, pp. 13-22.

Baldwin, Carliss Y. / Clark, Kim B. (2000): Design Rules – The Power of Modularity: Cambridge / Massachusetts: The MIT Press 2000.

Berman, Barry (2002): Should your firm adopt a mass customization strategy?, Business Horizons, Vol. 45, No. 4, pp. 51-60.

Blecker, Thorsten / Abdelkafi, Nizar / Kreutler, Gerold / Friedrich, Gerhard (2004): Product Configuration Systems: State of the Art, Conceptualization and Extensions, in: Abdelmajid Ben Hamadou / Faiez Gargouri / Mohamed Jmaiel (Eds.): Génie logiciel et Intelligence artificielle. Eight Maghrebian Conference on Software Engineering and Artificial Intelligence (MCSEAI 2004), Sousse / Tunisia, May 9-12, 2004, Tunis: Centre de Publication Universitaire 2004, pp. 25-36.

Blecker, Thorsten / Friedrich, Gerhard / Kaluza, Bernd / Abdelkafi, Nizar / Kreutler, Gerold (2005): Information and Management Systems for Product Customization, Boston et al.: Springer 2005.

Blecker, Thorsten / Friedrich Gerhard (Eds.): Mass Customization Information Systems in Business, Hershey/London: Idea Group Publishing 2006 (Forthcoming).

Byrne, Mike D. / Chutima, Parames (1997): Real-time operational control of an FMS with full routing flexibility, International Journal of Production Economics, Vol. 51, No. 1-2, pp. 109-113.

Chandra, Charu / Grabis, Janis (2004): Logistics and Supply Chain Management for Mass Customization, in: Charu Chandra / Ali Kamrani (Eds.): Mass Customization – A Supply Chain Approach, New York et al.: Kluwer Academic 2004, pp. 89-119.

Christopher, Martin (2005): Logistics and Supply Chain Management: Creating Value Adding Networks, 3rd Edition, Harlow et al.: Pearson Education Limited 2005.

Collier, David A. (1982): Aggregate Safety Stock Levels and Component Part Commonality, Management Science, Vol. 28, No. 11, pp. 1296-1303.

Coyle, John J. / Bardi, Edward J. / Langley Jr., C. John (2003): The Management of Business Logistics: A Supply Chain Perspective, 7th Edition, Ohio: South-Western 2003.

Da Silveira, Giovani / Borenstein, Denis / Fogliatto, Flávio S. (2001): Mass customization: Literature review and research directions, International Journal of Production Economics, Vol. 72, No. 1, pp. 1-13.

Doran, Desmond (2003): Supply chain implications of modularization, International Journal of Operations and Production Management, Vol. 23, No. 3, pp. 316-326.

Duray, Rebecca / Ward, Peter T. / Milligan, Glenn W. / Berry, William L. (2000): Approaches to mass customization: configurations and empirical validation, Journal of Operations Management, Vol. 18, No. 6, pp. 605-625.

Duray, Rebecca (2002): Mass customization origins: mass or custom manufacturing?, International Journal of Operations and Production Management, Vol. 22, No. 3, pp. 314-328.

Eynan, Amit / Rosenblatt, Meir J. (1996): Component commonality effects on inventory costs, IIE Transactions, Vol. 28, No. 2, pp. 93-104

Feare, Tom (2000): Pump the volume, Modern Materials Handling, Vol. 55, No. 3, pp. 55-59.

Fisher, Marshall L. / Ittner, Christopher D. (1999): The Impact of Product Variety on Automobile Assembly Operations: Empirical Evidence and Simulation Analysis, Management Science, Vol. 45, No. 6, pp. 771-786.

Forza, Cipriano / Salvador, Fabrizio (2002): Managing for variety in the order acquisition and fulfilment process: The contribution of product configuration systems, International Journal of Production Economics, Vol. 76, No. 1, pp. 87-98.

Garud, Raghu / Kumaraswamy, Arun (2003): Technological and Organizational Designs for Realizing Economies of Substitution, in: Raghu Garud / Arun Kumaraswamy / Richard N. Langlois (Eds.): Managing in the Modular Age – Architectures, Networks, and Organizations, Malden et al.: Blackwell Publishing 2003, pp. 45-77.

Guilabert, Margarita / Donthu, Naveen (2003): Mass Customization and Consumer Behavior: The Development of a Scale to Measure Customer Customization Sensitivity, Proceedings of the MCPC 2003, 2nd Interdisciplinary World Congress on Mass Customization and Personalization, Munich, October 6-8, 2003.

Gunasekaran, Angappa / Ngai, E.W.T. (2005): Build-to-order supply chain management: a literature review and framework for development, Journal of Operations Management, Vol. 23, No. 5, pp. 423-451.

Hart, Christopher W.L. (1995): Mass customization: conceptual underpinnings, opportunities and limits, International Journal of Service Industry Management, Vol. 6, No. 2, pp. 36-45.

Huffman, Cynthia / Kahn, Barbara E. (1998): Variety for Sale: Mass Customization or Mass Confusion, Journal of retailing, Vol. 74, No. 4, pp. 491-513.

Kärkkäinen, Mikko / Holmström, Jan (2002): Wireless product identification: enabler for handling efficiency, customisation and information sharing, Supply Chain Management: An International Journal, Vol. 7, No. 4, pp. 242-252.

Kotha, Suresh (1995): Mass Customization: Implementing the Emerging Paradigm for Competitive Advantage, Strategic Management Journal, Vol. 16, Special Issue, pp. 21-42.

Kotha, Suresh (1996): From mass production to mass customization: The case of the National Industrial Bicycle Company of Japan, European Management Journal, Vol. 14, No. 5, pp. 442-450.

Lampel, Joseph / Mintzberg, Hernry (1996): Customizing Customization, Sloan Management Review, Vol. 38, No. 1, pp. 21-30.

Langlois, Richard N. / Robertson, Paul L. (2003): Networks and Innovation in a Modular System: Lessons From the Microcomputer and Stereo Component industries, in: Raghu Garud / Arun Kumaraswamy / Richard N. Langlois (Eds.): Managing in the Modular Age – Architectures, Networks, and Organizations, Malden et al.: Blackwell Publishing 2003, pp. 78-113.

Lee, Hau (2004): The Triple-A Supply Chain, Harvard Business Review, Vol. 82, No. 10, pp. 102-113.

MacCarthy, Bart L. / Brabazon, Philip G. / Bramham, Johanna (2003): Fundamental modes of operation for mass customization, International Journal of Production Economics, Vol. 85, No. 3, pp. 289-304.

MacDuffie, John P. / Sethuraman, Kannan / Fisher, Marshall (1996): Product Variety and Manufacturing Performance: Evidence from International Automotive Assembly Plant Study, Management Science, Vol. 42, No. 3, pp. 350-369.

Meyer, Marc / Lehnerd, Alvin (1997): The Power of Product Platforms: Building Value and Cost Leadership, New York: The Free Press 1997.

Miller, George A. (1956): The Magical Number Seven, Plus or Minus Two: Some Limits on Our Capacity for Processing Information, The Psychological Review, Vol. 63, pp. 81-97, URL: http://www.well.com/user/smalin/miller.html (Retrieval: October 1, 2005).

Nilles, Volker (2002): Effiziente Gestaltung von Produktordnungssystemen – Eine theoretische und empirische Untersuchung, Ph.D. Dissertation, Technische Universitaet Muenchen, Muenchen 2002.

Perkins, A.G. (1994): Product variety: beyond black, Harvard Business Review, Vol. 72, No. 6, pp. 13-14.

Pine II, B. Joseph (1993): Mass Customization: The New Frontier in Business Competition, Boston, Massachusetts: Harvard Business School Press 1993.

Pine II, B. Joseph / Gilmore, James H. (1999): The Experience Economy, Boston, Massachusetts: Harvard Business School Press 1999.

Pine II, B. Joseph / Gilmore, James (2000): Satisfaction, sacrifice, surprise: three small steps create one giant leap into the experience economy, Strategy & Leadership, Vol. 28, No. 1, pp. 18-23.

Pine II, B. Joseph / Victor, Bart / Boynton, Andrew C. (1993): Making Mass Customization Work, Harvard Business Review, Vol. 71, No. 5, pp. 108-119.

Piller, Frank T. (1998): Kundenindividuelle Massenproduktion – Die Wettbewerbsstrategie der Zukunft, Muenchen, Wien: Hanser 1998.

Piller, Frank T. (2003): Mass Customization – Ein wettbewerbsstrategisches Konzept im Informationszeitalter, 3rd Edition Wiesbaden: Gabler Verlag 2000.

Piller, Frank T. / Ihl, Christoph (2002): Mythos Mass Customization: Buzzword oder praxisrelevante Wettbewerbsstrategie? – Warum viele Unternehmen trotz der Nutzenpotentiale Kundenindividueller Massenproduktion an der Umsetzung scheitern, Working paper, Department of Information, Organization and Management, Technische Universitaet Muenchen, URL: http://www.mass-customization.de/mythos.pdf (Retrieval: October 1, 2005).

Piller, Frank / Ihl, Christoph / Füller, Johann / Stotko, Christoph (2004): Toolkits for Open Innovation – The Case of Mobile Phone Games, in: Proceedings of the 37th Hawaii International Conference on System Sciences 2004.

Porter, Michael E. (1998): Competitive Advantage – Creating and Sustaining Superior Performance, 2nd Edition, New York et al.: The Free Press 1998.

Ramdas, Kamalini (2003): Managing Product Variety: An Integrative Review and research Directions, Production and Operations Management, Vol. 12, No. 1, pp. 79-101.

Riemer, Kai / Totz, Carsten (2001): The many faces of personalization – An integrative economic overview of mass customization and personalization, Proceedings of the MCPC 2001, 1st Interdisciplinary World Congress on Mass Customization and Personalization, Hong Kong, October 1-2, 2001.

Rogoll, Timm / Piller, Frank (2002): Konfigurationssysteme fuer Mass Customization und Variantenproduktion, Muenchen: ThinkConsult 2002.

Rosenberg, Otto (1996): Variantenfertigung, in: Werner Kern / Hans-Horst Schroeder / Juergen Weber (Eds.): Handwoerterbuch der Produktionswirtschaft, 2nd Edition, Stuttgart: Schaeffer-Poeschel 1996, pp. 2119-2129.

Sabin, Daniel / Weigel, Rainer (1998): Product Configuration Frameworks – A Survey, IEEE intelligent systems, Vol. 13, No. 4, pp. 42-49.

Sako, Mari / Murray, Fiona (1999): Modular strategies in cars and computers, Financial Times, No. 6, December.

Sonnenschein, Martin / Weiss, Michael (2005): Beyond mass customization: Initial thoughts of how to manage customer energy in competitive markets, in: Thorsten Blecker / Gerhard Friedrich (Eds.): Mass Customization: Concepts – Tools – Realization, Berlin: GITO-Verlag 2005, pp. 411-423.

Tersine, Richard J. / Wacker, John G. (2000): Customer-Aligned Inventory Strategies: Agility Maxims, International Journal of Agile Management Systems, Vol. 2, No. 2, pp. 114-120.

Toffler, Alvin (1980): The Third Wave, New York: William Morrow & Co., Inc. 1980.

Tseng, Mitchell, M./Jiao, Jianxin (2001): Mass Customization, in: Gavriel Salvendy (Ed.): Handbook of Industrial Engineering – Technology and Operations Management, 3rd Edition, New York et al.: John Wiley & Sons, INC. 2001, pp. 684-709.

Ulrich, Karl (1995): The role of product architecture in the manufacturing firm, Research Policy, Vol. 24, No. 3, pp. 419-440.

van Hoek, Remko I. (2000): Role of third party logistic services in customization through postponement, International Journal of Service Industry Management, Vol. 11, No. 4, pp. 374-387.

von Hippel, Eric (2001): User Toolkits for Innovation, Working paper, MIT Sloan School of Management, URL: http://ebusiness.mit.edu/research/papers/134%20vonhippel,%20Toolkits.pdf (Retrieval: October 1, 2005).

Wacker, John G. / Treleven, Mark (1986): Component Prt Standardization: An Analysis of Commonality Sources and Indices, Journal of Operations Management, Vol. 6, No. 2, pp. 219-244.

Wildemann, Horst (1995): Produktionscontrolling: Systemorientiertes Controlling schlanker Produktionsstrukturen, 2nd Edition, Muenchen: TCW Transfer-Centrum 1995.

Yeh, Kun-Huang/Chu, Chao-Hsien (1991): Adaptive Strategies for Coping with Product Variety Decisions, International Journal of Operations and Production Management, Vol. 11, No. 8, pp. 35-47.

Zipkin, Paul (2001): The limits of mass customization, Sloan Management Review, Vol. 42, No. 3, pp. 81-87.

Chapter 2

PERSONALIZED NEEDS ELICITATION IN WEB-BASED CONFIGURATION SYSTEMS

Gerold Kreutler and Dietmar Jannach
University of Klagenfurt,Institute for Business Informatics and Application Systems

Abstract: The high product variety of a mass customization strategy induces a high level of complexity both from the mass-customizer's perspective as well as from the customers' viewpoint. In particular, a high number of different product variants and configurable features can be challenging for the end-user who is often overwhelmed during the configuration and buying process. As customers are generally not technical engineers, but rather less-experienced, they are often confused and unable to choose the product that best fits their needs. As a consequence, customers can be dissatisfied with their buying decision later on, which finally leads to frustration and a decrease of customer loyalty. Web-based product configuration systems are nowadays well-established in commercial environments and enable users to specify desired product variants typically on a technical level. Thus, they efficiently support product experts in configuring their desired product variant. However, most current systems do not take into account the fact that online configuration systems should be usable and helpful for quite heterogeneous user groups. Online customers typically have a different background in terms of experience or skills or are simply different in the way they prefer to (are able to) express their needs and requirements. Thus, we argue that the typical "one-style-fits-all" approach for needs elicitation is not adequate for customer-supplier-interaction in mass customization. As users are different, it is necessary to adapt the interaction to the customer, i.e. to take the user's background or his capabilities into account and tailor the interaction accordingly. Within this paper, we comprehensively discuss personalization and adaptation possibilities for interactive needs elicitation in online configuration by categorizing the different levels and dimensions in a conceptual framework. Throughout, we describe adequate techniques for effectively implementing such functionality and give examples for personalization opportunities for the different levels. Finally, we discuss architectural aspects when building and maintaining such highly-adaptive web applications.Our work extends already existing work on personalization for product configuration systems. However, while most existing approaches base their adaptation features on long-term user models, we focus on (knowledge-based) techniques that allow us to personalize the interaction style also for first-time users, for which there is nearly no support in most existing systems.

Key words: Personalization, Web-based Configuration Systems, Needs Elicitation

1.   INTRODUCTION

Today, the competitive situation of companies is characterized by a strong orientation towards product individualization. The market's demand for customer-individual, configurable products has been constantly increasing. As a consequence, the mass-customization paradigm, which aims at satisfying individual customer needs with a near mass production efficiency (Pine, 1993), has been applied in different industrial sectors.

The high product variety of the mass customization strategy induces a high level of complexity both from the mass customizer's perspective as well as from the customer's view-point. Internal complexity induces additional (hidden) costs at the manufacturers' level, external complexity can lead to confusion during the customers' decision making process. In particular, the high number of different product variants and configurable features can be challenging for the end-user who is often overwhelmed during the configuration and buying process (Scheer et al., 2003). As customers are generally not technical engineers, but rather less-experienced, they are often unable to choose the product that best fits their needs. As a consequence, they can be dissatisfied with their buying decision later on, which finally leads to frustration and to a decrease of customer loyalty.

Web-based product configuration systems are important enablers of the mass customization paradigm and nowadays are well-established in commercial environments. They enable users to specify desired product variants – typically on a technical level, because in practice the technological perspective dominates the user perspective (Blecker et al., 2005). Thus, they efficiently support product experts in configuring their desired product variant. However, most current systems do not take into account the fact that online configuration systems should be usable and helpful for heterogeneous user groups. Online customers typically have a different background in terms of experience or skills or are simply different in the way they prefer to or are able to express their needs and requirements (Felfernig et al., 2002). Thus, we argue that the typical "one-style-fits-all" approach for needs elicitation, e.g. based on static HTML fill-out forms, is not adequate for customer-supplier-interaction in mass customization environments. As users are different, it is necessary to adapt the interaction process to the customer, i.e. to take the user's background or his capabilities into account and tailor the interaction accordingly. For example, if we think of a system for configuring personal computers, there will be users who want to specify technical details of the desired model, whereas others will only be able to express for what purposes they intend use the computer; others again only want to compare preconfigured models and decide by themselves.

In particular, the quality of the results, i.e. the accuracy of the acquired customers' real needs and consequently the proposed product configurations that best fit these needs, can be significantly improved when the system interacts with the user in a personalized way. Extensive personalization of the interaction between the user and the configuration system can bring us one step closer to real-world face-to-face communication where the communication partners adapt their communication style to their vis-à-vis. Thus, users are enabled to express their requirements in a natural way and their confidence in the system's results increases when they have the feeling that their requirements are taken adequately into account.

Within this paper, we comprehensively discuss personalization and adaptation possibilities for interactive needs elicitation in online configuration by categorizing the different levels and dimensions in a conceptual framework. Throughout, we describe adequate techniques for effectively implementing such functionality and give examples for personalization opportunities on different levels. Our work extends already existing work on personalization for web-based product configuration systems, e.g. Ardissono et al. (2003), introducing new personalization concepts that are already applied in web-based guided selling systems (see, e.g. Jannach (2004), Jannach and Kreutler (2004)). Whereas most existing approaches base their adaptation features on long-term user models, we focus on techniques for the personalization of the interaction also for first-time-users, for which there is nearly no support in existing systems up to now. Finally, we discuss architectural aspects for building and maintaining such highly-adaptive web applications.

2. PERSONALIZATION AND ADAPTATION IN THE CONFIGURATION PROCESS

Personalization can be considered as a means to help individuals satisfy a goal that efficiently and knowledgeably addresses their need in a given context by understanding their preferences (Ricken, 2000). In web-based e-commerce settings, personalization consists of activities that tailor the user's web experience to his or her particular needs, e.g. by adapting online applications to individual user's characteristics or usage behavior on several levels. In order to find a general classification scheme for the different personalization possibilities in the web-based configuration process, we follow the basic structure of Kobsa et al. (2001) who identified three basic categories of personalization opportunities for general hypermedia applications: Content level, interaction level, and presentation level. Note that a strict separation of these levels is not always possible, which may lead to overlaps in the categorization.

In contrast to existing work in this context, like for instance Ardissono et al. (2003), we do not primarily focus on the acquisition of a long-term user model, but rather on short-term personalization possibilities that can be immediately applied during an interaction, e.g. in the case of new users. Thus, it is possible to cope with the new-user-problem (Rashid et al., 2002). However, it is also possible to improve the presented concepts by the application of long-term user models that provide further information about the user.

2.1   Personalization on the Content Level

Configuration Steps and Configuration Dialog. A product configuration dialog typically consists of a set of subsequent questions about desired product features, i.e. the user is repeatedly asked to select or enter one or more values for a certain feature or option. The configuration engine uses these inputs to refine the current user's configuration, i.e. the product variant. This process is repeated until all required product features are selected. In non-adaptive approaches, every user is asked the same set of questions in the same order. However, this is problematic because it can lead to a configuration result that only poorly corresponds to the user's needs and preferences. The typical problems are, e.g.,

– the user does not understand a configuration step because of missing background knowledge. Thus, the user is unable to select some product features and the default value or even a wrong value is chosen.

– the user is annoyed by too many steps in the configuration process that are already irrelevant in the current situation due to previously given answers.

– the user is frustrated by a non-natural interaction style in which the system statically poses questions about product features without reacting situatively on the user's current answers.

Such situations do not only cause poor configurations that do not match the customers' requirements, they also reduce the user's confidence in the system's results, in particular if he has the feeling that he was not able to clearly express his needs.

In our approach, we aim at mitigating these problems by the application of personalization techniques during the configuration process on different levels. On the one hand, the presentation of the questions and the selectable features can be personalized, as well as the dialog flow between the user and the configuration system itself in order to achieve a more natural conversational interaction style (cf. Bridge (2002), Carenini et al. (2003)). Figure 2-1 depicts an overview of the personalization possibilities on a configuration page on the content level. The individual contents of the page can be dynamically constructed on the basis of a declarative knowledge base that contain the required text fragments as well as the personalization rules that determine the page content based on the current user's characteristics (Jannach, 2004).

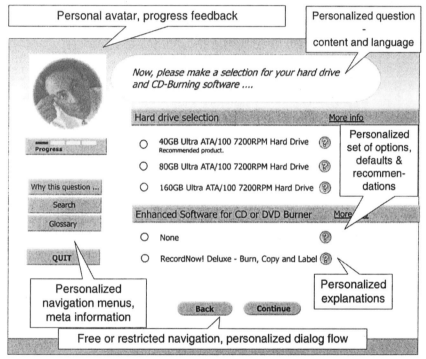

Figure 2-1. Personalized configuration page

In this context, the personalization capabilities comprise, for instance, the following items:

– The selection of a certain language or jargon that different user groups might be used to or feel most comfortable with, e.g. a formal or an entertaining language.

– The dynamic construction of the set of configuration features that can be selected by the user. This means that depending on the current state of the configuration, i.e. the user's previous inputs, some alternatives can be added or removed. Thus, irrelevant options (in the current situation) or too complex ones (for certain user groups) are removed.

– The automatic selection of appropriate situation-dependent defaults (system proposals) in order to minimize the number of required clicks, which is particularly important in longer dialogs.

– The amount of optional detailed information for a configuration step or options, depending on the user's estimated domain expertise.

On the level of the dialog itself, we propose a knowledge-based approach to design personalized user dialogs. Therefore, the web-based conversation can be modeled in terms of a sequence of configuration pages. These pages typically contain one or more questions where the user can set a product feature in his most convenient style. This comprises on the one hand the presentation, e.g. a product feature's graphical representation, on the other hand the content of the question. For instance, whereas product experts prefer to select product features directly, novice users are more familiar with customer-oriented questions about their needs where the configuration system then internally infers suitable product features.

In a knowledge-based approach, all the personalization rules, i.e. the selection of a configuration page and their contents in a certain situation, as well as the selection of a suitable presentation style or language, can be modeled as declarative conditions over the current user's characteristics. At run-time, the configuration system can automatically evaluate these conditions and choose appropriate configuration pages with a suitable presentation style. Note that the user characteristics to be evaluated can stem from already given answers of the user; in addition, also models of known users can be exploited.

Phases are a further means of personalization that structure the dialog. They can be used to provide the user some feedback on the dialog's progress and to vary the degrees of freedom with respect to navigation, i.e. whether a user is allowed to freely navigate between configuration steps.

Hints. The provision of optional opportunistic hints is another possibility of personalizing general hypermedia applications identified by Kobsa et al. (2001). Particularly in online product configuration, such hints are a major means to enrich the otherwise mostly system-driven dialogs because they give an immediate and personalized feedback on the user's inputs. Consequently, users get the feeling that the system actively monitors their inputs and participates in the dialog. Hints are applicable on different levels and can be again modeled as conditions over the current user's characteristics.

The major benefit of hints is to provide additional information about certain configuration options, i.e. product features that can be set by the user. Thus, it is possible to provide non-expert users with detailed technical information or to display additional information for cross-selling or up-selling purposes. Additionally, hints can be used to actively interrupt the dialog, in particular in cases where the user has to be informed about possible inconsistencies in his requirements that lead to an empty configuration result. Finally, they allow for the personalization of the result page, i.e. the last dialog step presenting the configured product proposals. There, it is possible to provide supplementary information on the displayed configurations. Furthermore, it is possible to explain additional inferences on the user requirements in cases when the system applied internal reasoning rules to infer user preferences that cannot be directly acquired.

Explanations and Reasoning. The results of the product configuration process are valid product configurations that correspond to the customer's real needs and preferences. In order to increase the user's confidence into the system's output, the system has to provide understandable explanations. We argue that these explanations also have to be personalized to be understandable and useful for different kind end-users. Depending on the current user's capabilities and interests, several points can be varied, such as:

– The language used in the explanations (e.g. technical or non-technical terms).

– The level of details of the underlying reasoning process that are presented, i.e. information provided by the configuration engine.

A specific form of personalization of the reasoning process is to enable the user to override the outcome of the reasoning process to some extent (Jannach and Kreutler, 2005). In online configuration systems, a typical example are indirectly acquired (derived) customer characteristics where the system infers some estimate of customer properties that cannot be acquired directly, e.g. the risk class of a customer in an investment scenario. The further reasoning process is then based on the outcome of that classification which should also be part of the explanation the system provides. Moreover, enabling the (advanced) user to override these estimates can also lead to a more accurate elicitation of the user's personal needs and better configuration results.

Finally, the personalization of the configuration system's reasoning behavior can also influence the treatment of unsolvable user requirements. This means that the configuration system is not able to find a valid product configuration that corresponds to the preferences and requirements of the user, i.e. to his/her inputs. In these situations, the system has to remove some user constraints in order to find a valid product that fulfills as many requirements as possible (see, for instance, Freuder and Wallace (1992)). Therefore, the user has to state priorities for his requirements, which consequently increases the accuracy of the estimate of his/her interests. In the explanation phase, the system can then use the lists with all requirements that were fulfilled as well as the requirements that were dropped.

Result Presentation. Even in the phase when the suitable product configurations are presented, there are some personalization opportunities to enable the user to refine his requirements.

One possible option is the presentation of alternative products, both relatively similar ones as well as reference products from other classes of products. Thus, the application of different similarity measures and the provision of adequate explanations, e.g. how a product fits the user's stated requirements or not, lead to a more accurate user model.

In general, the result presentation phase can be used to monitor the quality of the configuration process over time. This can be done for instance by letting the users submit ratings whether he found the proposal useful or not, or by monitoring the click-behavior of the user (e.g. for clicking on a link for viewing detailed product information).

2.2   Personalization on the Interaction and Presentation Level

On the interaction level, two aspects of personalization can be considered: the interaction style and degrees of freedom in navigation. Regarding the interaction style, in online configuration systems one basic form is common – a system driven dialog with fill-out forms (which can be extended by extensive personalization through a dynamically adapted front-end). Most importantly, online users are well-acquainted with this interaction style; they also often feel comfortable when the system actively guides them through the configuration process. Nonetheless, depending on the current user and on the application domain, other forms of interaction can be more intuitive for the user and finally lead to better results in the elicitation process.

Natural Language Interaction. In this context, the most important personalization aspect is the decision how "user driven" a dialog should be designed, i.e. whether the user should be enabled to actively steer the dialog, e.g. by directly posing questions. In theory, the ultimate solution for this would be a full natural-language interface based on an intelligent agent that has both the knowledge in the application domain as well as the required knowledge to carry out a conversation, i.e. how to steer the dialog or react to specific situations. First natural-language style approaches are already applicable in e-commerce settings (see, e.g., Thiel et al. (2002), Thompson et al. (2004)), but there are still open problems. Particularly the requirement of massive knowledge acquisition and modeling efforts to reach an acceptable dialog quality is a restraint for the implementation of such a system. Most importantly, it is difficult to cope with general user utterances beside from domain knowledge. This could be interpreted as a "poor" dialog quality by the user, which consequently leads to a frustration because users attribute more intelligence to the system than there actually is. Furthermore, in many application domains the dialog cannot be fully user-driven because the user's background knowledge is too limited (i.e. he cannot properly articulate questions).

Degrees of Freedom in Navigation. Another way to vary the interaction style according to the current user's needs and capabilities is the variation of the "degrees of freedom" with respect to navigation. This refers to the guidance of the users in the dialog. whereas some users might prefer a strong guidance, i.e. a strict order of configuration steps, others feel more comfortable when they can steer the dialog on their own. This comprises the possibility of selecting the order of questions they answer, moving forth and back in the dialog, revising answers, or trying different alternatives for product features. Additionally, the amount of visible navigation functionality for the user can be personalized, such that experts do not feel restricted in their possibilities, while beginners are not overwhelmed or frustrated by the complexity of the application.

Domain-specific Interaction Styles. In state-of-the-art product configuration applications, users have to specify the details of the desired configuration by going through a guided dialog where they have to answer several questions about desired product features and/or their preferences. In real life, however, customers are not tied to one single style in human-human-interaction. They prefer different communication styles with their vis-à-vis, depending on the current situation and the domain they are in. For instance, in the financial domain, clients are used to be presented a product proposal from their sales person after an intensive requirement elicitation dialog, whereas in domains of consumer goods, e.g. digital cameras, expert customers could expect support from the sales person in comparing several products.

Therefore, in the online channel customers also must not be constrained to one single interaction style. Depending on the current user's situation, personalized online configuration systems also should offer several interaction styles. Besides the described standard dialog that leads to product configurations, some users could prefer to start with a basic, pre-configured model and adapt one or the other part; others again only want to specify some key components and functionality and let the system decide on the rest. The selection of the appropriate interaction style can be done either explicitly by the user at the beginning of the configuration process, or implicitly by the system, e.g. by asking the user a few questions to determine the most suitable interaction form.

Domain-specific interaction styles enable users to express their requirement in several ways. Therefore, we argue that the overall quality of the results of the online configuration process also increases.

Presentation Style. Kobsa et al. (2001) identified the presentation level as the third level of personalization. In our context, this level is strongly related with the interaction and content levels. In general, all "standard" personalization possibilities as described by Kobsa et al. (2001) can be applied. This comprises, e.g., support for different end-devices or handicapped users by different font-sizes or adaptable contrast. In the special context of online product configuration, personalized presentation variants could be provided with respect to the following dimensions.

First, the configuration dialog can be executed in an own window that focuses the user on a small area of interest, or integrated in a surrounding website or portal. There, additional information like, e.g. glossaries, further links or frequently asked questions), can be easily incorporated, which is advantageous for users that actively search for more information during the configuration process.

Another form of personalization of the presentation can be an appropriate interface layout that is coordinated with the language style used in the configuration process. For instance, a less formal or entertaining language can be supported by an animated "avatar" that serves as virtual conversation partner. This livens up a guided dialog and increases the user's online experience.

3.   ARCHITECTURAL REQUIREMENTS

There are some major challenges involved in the development of extensively personalized web applications: Personalization is known to be a knowledge-intensive task (Kobsa et al., 2001). Such systems therefore have to feature adequate means for acquiring, representing, and – in particular – maintaining the required personalization knowledge. In addition, personalized user interfaces have to be extremely flexible, because both the content as well as the navigation options have to be dynamically determined and displayed based on the underlying personalization rules. Typically, there are also strong interdependencies between user interface, reasoning, and the knowledge base, which are challenging from an engineering perspective, because a clear separation between the application components in the sense of the Model-View-Controller approach (Krasner and Pope, 1988) can be difficult.

In Figure 2-2 we give an overview of a possible architecture for a personalized configuration service (compare, e.g., Jannach and Kreutler (2005)). One of the major features of this architecture is that we propose having as much as possible of the required knowledge in a shared repository. In particular we argue that the knowledge representation mechanism needed for expressing e.g., configuration and personalization knowledge should be based on a shared conceptualization and on compatible problem solving techniques as much as possible. Note that a different approach was taken e.g., in the CAWICOMS (Ardissono et al., 2003) project, where the core configuration task was based in Constraint Satisfaction, whereas personalization was based in rules and dynamic evaluation of user preferences.

In addition, also the required the knowledge acquisition and maintenance tools have to be integrated in a way that the knowledge engineer can edit the different pieces of knowledge in a consistent way, e.g., by using the same sort of "constraint language" for expressing configuration and personalization rules.

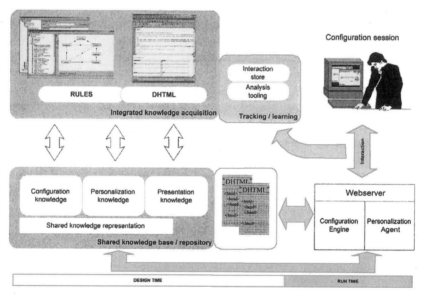

Figure 2-2. Possible architecture for a personalized configuration service

In many applications, the hardest part can be the integration of the development of the dynamic web pages: First, we have to deal with the limitations and shortcomings of dynamic HTML. In addition, we also have to take into account that – although most of the personalized content has to be generated dynamically – the pages have to be maintainable by Web developers that e.g., adapt the layout according to a company's web site. A possible solution for the domain of personalized, content-based product recommendation was presented by Jannach (2004), where the dynamic web pages are constructed from modular page fragments and JSP "Custom Tags" where used to hide the complexity from the Web developer.

At run-time, our architecture proposes two modules, a configuration engine and a personalization agent that manages the interaction with the end user. Of course, both engines can make use of the same underlying problem reasoner, e.g., a constraint solver.

A final part of the architecture is "tracking/learning". Depending on the used personalization mechanisms it can be possible that the system fine-tunes itself over time (e.g., when using a sort of Multi-Attribute-Utility mechanism, or that a knowledge-engineer can do offline analyses and manually adapts/corrects the personalization rules, which is needed in many personalized systems after the initial setup.

4.   SELECTED EXAMPLES FOR SUCCESSFUL PERSONALIZATION TECHNIQUES

Up to now, personalized needs acquisition has not been an important research issue in the field of configuration systems. Instead, personalizing the preference elicitation process was primarily addressed in the context of recommender systems. In this chapter, we describe some deliberately selected approaches that describe already established personalization techniques in the field of recommender and configuration systems.

McGinty and Smyth (2002a) propose an approach for recommender systems that is based on a more casual conversation. This means that there should be several degrees of feedback that an online user can provide during the dialog. For instance, leading users through deep dialogs that replicate customer buying models from real world is not appropriate in most online recommender settings. In such dialogs, users are asked direct questions about important product features, as real-world sales assistants would do in some cases. However, online users are less tolerant of being asked detailed questions, which prevents them from entering such dialogs.

Therefore, the authors argue that there should also be a low-cost form of feedback for users and propose a comparison-based recommendation approach, in which the user is asked to choose a recommended item as a (positive or negative) preference. The further product recommendation is based on the difference of the preferred products and the remaining alternatives.

In further work, McGinty and Smith (2002b) give an overview on different techniques for user feedback, e.g. value elicitation, tweaking, ratings-based, and preference-based. In their paper, they focus on a low-cost preference-based feedback model which is evaluated in a recommendation framework.

The proposed feedback techniques for recommender systems can also be applied in the context of configuration systems. For instance, in the result presentation phase comparison-based approaches can be applied. In the sense of our work, different feedback techniques are a suitable means for the personalization on the interaction level and are comparable to domain-specific interaction styles.

Shimazu (2001) proposes the agent system ExpertClerk that imitates a human salesclerk that supports customers in finding the suitable product in online stores. The system supports two basic techniques: *Navigation by asking* and *navigation by proposing*. First, the system carries out a natural-language conversation with the customer in order to find a set of suitable products. Then, after a pre-defined threshold of questions, the three most-contrasting products among the remaining products are compared. Both steps are repeated until an appropriate product is found.

When compared with our work, the proposed system applies techniques on the interaction layer. It is noteworthy that ExpertClerk alternately takes use of different interaction styles during one dialog with the user, which enables the system to situively react on the current state of the interaction. The system also depicts the current state of the dialog, i.e. it is show how many products are left for recommendation. Thus, the user can not get lost during the elicitation process.

The ClixSmart Navigator architecture (Smyth and Cotter, 2002) introduces personalization capabilities for mobile portals on the interaction level. It supports users navigating to the content of their interests in WAP portals. The authors identify that excessive navigation times, e.g. for navigating through a series of menu, frustrate users and are jointly responsible for the little success of WAP portals. Therefore, the ClixSmart Navigator adapts the structure of a mobile portal to the personal needs of users by storing hit tables which track an individual user's navigation behavior. Based on these tables, the menu is adapted (i.e. the position of menu items is reordered) to minimize the navigation distance for the most probable navigation options for a user.

Although the field of the proposed architecture is quite different from the product configuration domain, the main idea can also be applied for configuration systems: It is essential that the dialog with the end-user is carried out in a personalized way. This means that the user must not be annoyed with configuration steps that are not useful in the current situation of the dialog. Thus, it should be possible to shorten the dialog by dynamically selecting only relevant configuration steps.

In the domain of configuration systems, Pu et al. (2003) consider preference elicitation as a fundamental problem. Stemming from experiences in building decision support systems in various domains, they identify some principles for designing of the interactive procedure of finding a solution, i.e. a suitable configuration in the solution space.

In a survey of 10 commercial online flight reservation systems, they find out that a personalized order elicitation improves the preference elicitation process for the end-user. Thus, users should be able to state values for those options that correspond to their main objectives, which leads more quickly to a more accurate preference model. Furthermore, example critiquing in a minimal context, i.e. making critiques on a personalized (minimized) set of attributes, is also identified as adequate means.

The authors also consider the visualization of the result set with the possibility of revising previously stated preferences during the elicitation process as crucial because users can immediately see the consequences of their stated preferences and possible changes.

5.   CONCLUSIONS

Nowadays, web-based configuration systems are well-established in industrial environments and essential for the success of the mass customization paradigm. However, state-of-the-art configuration systems are mainly product-oriented and do not optimally support heterogeneous groups of end-users in the configuration process, which often overwhelms customers and leads to frustration. In this paper, we have argued that personalization is a key factor to hide the external complexity and elicit the customer's real needs to lead him successfully to a suitable product configuration. Therefore, we have given an overview on personalization in this context and made a conceptualization of the personalization possibilities. Throughout, we focused on techniques that are also applicable for first-time users where no long-term user model exists. Finally, we have presented some architectural aspects for the development and maintenance of such extensively personalized web applications.

ACKNOWLEDGMENTS

This work is partly funded by grants from the Austrian Central Bank, OeNB, No. 9706, and from the Kärntner Wirtschaftsförderungsfonds.

REFERENCES

Ardissono L., Felfernig A., Friedrich G., Goy A., Jannach D., Petrone G., Schäfer R., Zanker M. 2003. A Framework for the Development of Personalized, Distributed Web-Based Configuration Systems. AI Magazine, 24(3), 93-110.

Blecker T., Friedrich G., Kaluza B., Abdelkafi N., Kreutler G. 2005. Information and Management Systems for Product Customization. Springer, New York.

Bridge D. 2002. Towards Conversational Recommender Systems: A Dialogue Grammar Approach. Proc. of the Workshop in Mixed-Initiative Case-Based Reasoning, Workshop Programme of the 6th European Conference in Case-Based Reasoning, 9-22.

Carenini G., Smith J., Poole D. 2003. Towards more Conversational and Collaborative Recommender Systems. Proc. of the Intelligent User Interfaces 2003 (IUI'03), Miami, USA, 12-18.

Felfernig A., Friedrich G., Jannach D., Zanker M. 2002. Web-based configuration of virtual private networks with multiple suppliers. Proc. of the Seventh Intl. Conference on Artificial Intelligence in Design (AID'02), Cambridge, UK, 41-62.

Freuder E.C., Wallace R.J. 1992. Partial constraint satisfaction. Artificial Intelligence, 58, Issue 1-3, 21-70.

Jannach D. 2004. Advisor Suite – A knowledge-based sales advisory system. Proc. of the 16th European Conference on Artificial Intelligence (ECAI2004), 720-724.

Jannach D., Kreutler G. 2004. Building On-line Sales Assistance Systems with ADVISOR SUITE. Proc. of the Sixteenth International Conference on Software Engineering & Knowledge Engineering (SEKE 2004), Alberta Banff, Canada, June 2004, 110-117.

Jannach D., Kreutler G. 2005. Personalized User Preference Elicitation for e-Services. Proc. of the 2005 IEEE International Conference on e-Technology, e-Commerce, and e-Service, Hong Kong, 2005.

Kobsa A., Koenemann J., Pohl W. 2001. Personalized Hypermedia Presentation Techniques for Improving Online Customer Relationships. The Knowledge Engineering Review, 16(2), 11-155.

Krasner G.E., Pope S.T.. 1998. A Description of the Model-View-Controller User Interface Paradigm in the Smalltalk-80 System. ParcPlace Systems Inc., Mountain View.

McGinty L., Smyth B. (2002a). Deep Dialogue vs Casual Conversation in Recommender Systems. In: Ricci F., Smyth, B. (Eds). Proceedings of the Workshop on Personalization in eCommerce at the Second International Conference on Adaptive Hypermedia and Web-Based Systems (AH-02), Universidad de Malaga, Malaga, Spain, 80-89.

McGinty L., Smyth B. (2002b). Comparison-Based Recommendation. In: Lecture Notes of Computer Science 2416, Proceedings of the 6th European Advances in Case-Based Reasoning, 575-589.

J. Pine II J. 1993. Mass Customization: The New Frontier in Business Competition. Harvard Business School Press, Boston.

Pu P., Faltings B., Torrens M. 2003. User-Involved Preference Elicitation. In: Eighteenth International Joint Conference on Artificial Intelligence (IJCAI'03), Workshop on Configuration, Acapulco.

Rashid A.M., Albert I., Cosley D., Lam S.K., McLee S.M., Konstan J.A., Riedl J. 2002. Getting to Know You: Learning New User Preferences in Recommender Systems. Proc. of the Intelligent User Interfaces 2002 (IUI'02), San Francisco, USA, 127-134.

Ricken D. 2000. Introduction: personalized views of personalization. Communications of the ACM. Special Issue on Personalization, 43(8), 26-28.

Scheer C., Hansen T., Loos P. 2003. Erweiterung von Produktkonfiguratoren im Electronic Commerce um eine Beratungskomponente. Working Papers of the Research Group Information Systems & Management, Johannes Gutenberg-University Mainz.

Shimazu H. (2001). ExpertClerk: Navigating Shoppers' Buying Process with the Combination of Asking and Proposing. In: Seventeenth International Joint Conference on Artificial Intelligence (IJCAI'01), Seattle, USA, 1443-1450.

Smyth B., Cotter P. (2002). Personalized Adaptive Navigation for Mobile Portals. In:. Proceedings of the 15th European Conference on Artificial Intelligence – Prestigious Applications of Intelligent Systems (PAIS), Lyons, France.

Thiel U., Abbate M.L., Paradiso A., Stein A., Semeraro G., Abbattista F. 2002. Intelligent E-commerce with guiding agents based on Personalized Interaction Tools. In: Gasos J., Thoben K.-D. (Eds.). E-Business Applications, Springer, 61-76.

Thompson C.A., Göker M.H., and Langley P. 2004. A Personalized System for Conversational Recommendations. Journal of Artificial Intelligence Research, 21, 393-428.

Chapter 3

MODEL-BASED CONFIGURATION SUPPORT FOR SOFTWARE PRODUCT FAMILIES

Katharina Wolter[1], Lothar Hotz[2] and Thorsten Krebs[1]
[1]University of Hamburg, Department of Informatics; [2]HITeC e.V., University of Hamburg

Abstract: In this paper, we present main aspects of the ConIPF methodology which can be used to derive customer-specific software products. The methodology is based on software product families and model-based configuration. First results from using the methodology in an industrial context are presented.

Key words: Product Derivation; Software Product Families; Application Engineering; Model-based Configuration; Software Configuration.

1.   INTRODUCTION

One approach used for software mass customization are software product families. They provide a highly successful approach for strategic reuse of product components (Parnas 1976). However, one major problem within family-based software engineering is the lack of methodological support for application engineering. The large number of decisions and dependencies between these decisions make the task complex and error-prone. Impacts of decisions are not known or overseen during application engineering. Functionality is implemented anew where reuse would have been possible because the large number of artifacts is hardly manageable. The methodology described in this paper combines the well known research areas of software product families and model-based configuration in order to fill this gap.

The methodology is based on a configuration model that represents functionality and variability provided by the product family. Basically, the configuration model provides two layers of configurable assets, i.e. a feature layer and an artifact layer. The artifact layer reflects the (variable) structure of the product family artifacts and the feature layer is the customer view on the functionality in the artifact layer. A mapping between the feature layer and the artifact layer allows for automatically inferring the needed software artifacts for a given selection of features. We call the resulting process of application engineering enhanced with automated inferences *model-based product derivation.*

In an ideal product derivation process the features required by the customer are selected and the configuration tool automatically infers all artifacts needed to provide these features. However, requirements may not be accounted for in the shared product family artifacts and can only be accommodated by adaptation or even new development. This involves adapting the product (family) architecture and / or adapting or creating component implementations. Since the software artifacts are represented in the configuration model, these modifications have to be aligned in a synchronized fashion. To support this, we developed a dependency analysis that can be performed on the configuration model.

Introducing the methodology in selected industrial environments dealing with software product families shows that it is applicable and can be tailored to the specific needs of particular organizations. The first results are promising and are presented later in the paper.

The remainder of this paper is organized as follows: in Section 2, we explain the relation between mass customization and model-based configuration. Subsequently, we introduce the approaches our methodology is based on (Section 3). Furthermore, we present the configuration tools KONWERK and EngCon, which we used to implement the methodology at our industrial partners. One of these application domains is described in Section 4. This domain is used as a guiding example. Section 5 introduces the methodology in detail by describing the configuration model, the product derivation process, and aspects of adapting existing and developing new components. In Section 6, we present experiments performed to validate the methodology and our experiences so far. In Section 7, we discuss related work and give a conclusion in Section 8.

2.   MASS CUSTOMIZATION VS. CONFIGURATION

The topics of model-based configuration from the area of artificial intelligence and mass customization have a lot in common. To analyze their relation more closely, definitions for the two terms are needed. *Model-based configuration* supports the composition of (technical) products from individual parameterizable objects to a *configuration* that fulfill a certain task (or purpose) (Günter 1995). Model-based configuration will be introduced in more detail in Section 3.2. For the term *mass customization* several different definitions are used as Piller (2003) states. Davis refers to mass customization when "the same large number of customers can be reached as in mass markets of the industrial economy, and simultaneously they can be treated individually as in the customized markets of pre-industrial economies" (1987, p. 169). The commonality between the two research areas is the *customer-specific product*. Model-based configuration support can be used to derive such customized products. However, the method is not restricted to mass market products. It is also applied successfully for complex capital goods (aircraft, drive systems, etc.).

According to Günter (2002) three different sales scenarios can be distinguished (first published in German in Günter (2001)):

- *Click & Buy*: Only non-customizable products are offered and thus no configuration support is needed in order to ensure the consistency of products.

- *Customize & Buy:* Products can be customized but the number of valid combinations of components is still restricted. Thus, complexity and consistency problems still play a minor role.

- *Configure & Buy:* A vast number of possible combinations of components and complex restrictions lead to a serious complexity problem. In this scenario it is not possible to list all possible products in a catalog. Model-based configuration support is needed for assembling valid products.

Similar categories are described by O (2002).

The methodology described in this paper does not fit to one of the above-introduced scenarios. Even in the last scenario a product can only be assembled from the components modeled so far. It is not possible to meet customer requirements, which have not been taken into account during developing the product family. In order to meet "new" customer requirements it might be necessary to adapt existing or develop new components – i.e. leaving the area of traditional routine configuration. Our methodology integrates these developing steps and the traditional product configuration process more closely. To complete the above described set of scenarios we add the following:

- *Configure, Develop & Buy:* This allows for assembling a product from components in the asset store and may also include components not yet developed. The main benefit is that it is possible to meet customer requirements that have not been taken into account yet. In addition to the complexity problem described for Configure & Buy, in this scenario it is necessary to integrate the new components and their relations into the existing configuration model. The facilities of routine configuration techniques are not sufficient to fulfill this task.

With this scenario, we move a step towards truly customized products. However, since the newly developed components are integrated into the product family they can be efficiently reused for all future customers.

The first three scenarios mentioned above are mainly applied to hardware products. However, customization is an important topic not only for hardware products but also for software products or software-intensive systems consisting of both hardware and software. On the one hand, software systems (must) become larger and of a higher quality because of increasing customer requirements and more complex system functionality. On the other hand, there is a need for reducing costs and shortening time-to-market in order to stay competitive. Offering more and more functionality inevitably leads to the Configure & Buy scenario. Selling functionality not yet implemented (e.g. because of new customer requirements) leads to Configure, Develop & Buy. The methodology described in this paper supports this last scenario.

3.   BASIC TECHNOLOGIES

In this section, we briefly introduce the approaches our methodology is based on: i.e. software product families (Section 3.1) and model-based configuration (Section 3.2). Additionally, in Section 3.3 we present model-based configuration tools used to implement and validate the methodology described in this paper.

3.1   Software Product Families

In software product families, the development of a product line and the development of products can be distinguished. These engineering tasks are identified as *domain engineering* and *application engineering* (compare Bosch et al. 2001):

- In *domain engineering*, architectures and reusable software components are developed. Exploiting commonality and managing variability is necessary and can be achieved by using feature models (Kang et al. 2002). Features are 'prominent or distinctive user-visible aspects of a system' (Kang et al. 1990) and can be modeled in partonomies with mandatory, optional and alternative properties.
- In *application engineering* existing artifacts are used to assemble specific products by analyzing requested features, selecting architecture and adapting components.

Domain engineering and application engineering do not describe chronological tasks, but the distinction between developing a product line and developing products using the product line. Application engineering is currently realized by communication facilities between developers and with standardized documents. These documents are used to capture customer requirements, to define system specifications and to realize change management. However, a general methodology for realizing or supporting application engineering does not exist (Hein et al. 2003). Therefore, it is common to use previously developed products or platforms by a "copy and modify approach" to suit current customer needs. This is rather error-prone in the sense that functionality is implemented anew where reuse would have been possible or incompatibilities between code fragments may not be detected leading to incorrect solutions.

3.2   Model-based Configuration

Configuration is a well known approach to support the composition of products from several parts. The configuration of technical systems is one of the most successful application areas of knowledge-based systems (Günter and Kühn 1999). In model-based configuration, basic modeling facilities enable the differentiation between three kinds of knowledge:

- *Conceptual knowledge* includes concepts, taxonomic and compositional relations as well as restrictions between arbitrary concepts (constraints).
- *Procedural knowledge* declaratively describes the configuration process.
- A *task specification* defines properties and constraints known from the customer that the product must fulfill.

The configuration process itself is performed in an incremental approach, where each step represents a configuration decision and possibly includes testing, simulating or checking with constraint techniques (Günter 1995, Hotz et al. 2003). However, applying configuration methods to software systems is in an early stage. First approaches are e.g. described in Soininen et al. (1998).

3.3   Tools

Two configuration tools have been used to implement, validate and improve the methodology: KONWERK and EngCon. KONWERK is a configuration tool developed partly at the University of Hamburg. EngCon is a scalable and flexible configuration platform from encoway GmbH[1]. EngCon was used in the project ConIPF to implement and validate the product derivation methodology at both industrial partners (see also Section 6). It was also extended with new functionality that has been discovered as necessary

---

[1]   www.encoway.de

for deriving software products. KONWERK was used to identify and per-petuate further research topics. Both tools are briefly introduced in the fol-lowing:

- *KONWERK* is a kernel system for configuration tasks (Günter and Hotz 1999). KONWERK is a modular system intended to make different areas of knowledge-based configuration accessible (like optimization, spatial configuration, cased-based configuration and support for vague model-ing). Thus, KONWERK provides a general architecture of knowledge-based systems for configuration and construction of technical systems. The implementation was mainly done by using Common Lisp, Common Lisp Object Systems (CLOS) and its meta-object protocol (see Kiczales et al., 1991). The user interface was implemented with the Common Lisp Interface Manager (CLIM). Thus, it is portable and runs under Windows and SunOS.

- *EngCon* is methodologically based on KONWERK. But in detail there are various differences and extensions, as e.g. further described in Holl-mann et al. (2000). EngCon has a flexible component architecture based on Java technology. Encoway provides a web-based modeling environ-ment (called K-Build), a powerful graphical GUI builder (K-Design) and development and connectivity tools (e.g. for SAP and other ERP, PDM and CRM systems) for the flexible customization of a configuration ap-plication. The configuration platform EngCon can be used in offline and online configuration scenarios. The user can configure either user-con-trolled or by using a configuration wizard (Ranze et al. 2002).

4.   APPLICATION DOMAIN

In this section we introduce the application domain used as a guiding ex-ample throughout the following sections of this paper. We implemented and tested the methodology at Robert Bosch GmbH[2] in the research unit Automotive Electronics which develops Car Periphery Supervision systems.

Car Periphery Supervision (CPS) systems monitor the local environment of a car. CPS systems are automotive systems based on sensors installed on a vehicle. The recording and evaluation of sensor data enable different kinds of high-level applications that can be grouped into safety-related and com-fort-related applications (see Thiel et al., 2001).

Two examples are given in the following:

- *Pre-Crash Detection (PCD)*: Based on sensor data it is possible to esti-mate the time, area and direction of an impact before the crash happens.

---

[2] www.bosch.de

This enables adjusting trigger points of specific airbags (Pre Set) in different locations in the car and firing a (seat) belt tensioner (Pre Fire) appropriately for the estimated crash situation.

- *Parking Assistance (PA)* supports the driver in avoiding moving the vehicle against people or stationary objects while driving slowly backward or forward. This is especially useful for vehicles that are difficult to look over.

5. THE METHODOLOGY

After introducing the basic approaches our methodology is based on and the application domain of one of our industrial partners, in this section, we describe the main parts of our product derivation methodology. The methodology as a whole is described in (Hotz et al. 2005).

The methodology is based on a configuration model where configurable assets of different types and their relations are formalized. This configuration model is introduced in Section 5.1. The model-based product derivation process is explained in Section 5.2. In Section 5.3 we address how necessary modifications for existing assets and new development can be combined with product derivation. All aspects are presented in general and with an illustrating example using the configuration tool EngCon.

5.1 Configuration Model

Model-based configuration provides means for modeling configurable assets and reasoning with them. Traditionally these are *hardware artifacts*. When considering software product families and software-intensive systems, also *software artifacts* and *features* need to be modeled. Additionally, we identified that the *context* in which the software-intensive system will be used is of particular importance for software-intensive systems. The context can influence the set of possible solutions although it is not part of the system itself. In the CPS domain, for example, some systems cannot be used in Europe or Northern America because of legal regulations or weather conditions of some areas restrict the choice of hardware artifacts.

Since these four asset types (features, context, hardware and software artifacts) are common to most application domains of software-intensive systems we defined them and their relations in a Commonly Applicable Model (CAM). A product, i.e. the result of the product derivation, contains software and hardware artifacts as parts and these together realize certain features. The Common Applicable Model has been further developed in the final version of the ConIPF methodology. A language called Asset Modeling for

Product Lines (AMPL) has been defined for which a metamodel (AMPL-M) similar to the CAM exists. For more details on this extension see (Hotz et al. 2005).

To implement the methodology in a specific domain the CAM is extended with domain-specific knowledge about hardware and software artifacts, the existing features and so on. An example from the CPS domain introduced in Section 4 is given in Figure 3-1. Please note, that the example only illustrates a part of the CAM and the domain-specific knowledge.

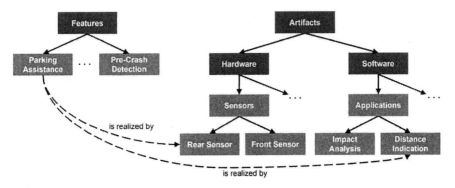

Figure 3-1. Example for the definition of domain-specific knowledge based on the CAM. The CAM asset types (`feature`, `hardware` and `software` `artifacts`) are emphasized in dark boxes while the domain-specific entities are distinguished by light-colored boxes

The CAM and the domain specific assets are defined in the *configuration model* as *concepts*. A clear separation of potential features and artifacts of the product family on the one side and features and artifacts that are used for a specific product derivation on the other side is needed. In model-based configuration tools this is realized by distinguishing between *concepts* for describing features and artifacts in general and *concept instances* for describing features and artifacts chosen for a specific product.

In the following we give a description of how concepts are defined in the model-based configuration approach:

- Each concept has a *name* which represents a uniquely identifiable character string. An example from the CPS domain is the concept called Parking Assistance.
- Concepts are related to other concepts in the *taxonomic relation*. A concept has exactly one *superconcept* and possibly several subconcepts. The concept Parking Assistance has the superconcept Feature which is part of the CAM.
- Attributes of concepts can be represented through *parameters*. A parameter is a tuple consisting of a name and a value descriptor. Diverse types of domains are predefined for value descriptors – e.g. integers, floats,

strings, sets and ranges. The concept Parking Assistance has one parameter called Symmetry with two possible values: Passenger Side or Passenger and Driver Side.

- Partonomies are generated by modeling *compositional relations*. Such a relation definition contains a list of parts (i.e. other concept definitions) that are identified by their names. Each of these parts is assigned with a minimum and a maximum cardinality that together specify how many instances of these concepts can be instantiated as parts of the aggregate. The CAM defines different compositional relations, e.g. has Features, has Hardware, has Software to emphasize on the difference between artifacts and features.

Both, the taxonomic relation and the compositional relation define a hierarchical structure between concepts. In addition to these relations, further relations can be used that do not define such a hierarchical structure. Examples are the *requires* relation and the *excludes* relation which can be used between arbitrary concepts. A further example is the *realizes* relation (*is realized by*). This relation is defined between features and artifacts only and expresses that features are realized by artifacts in an *n*-to-*m* mapping: one feature can be realized by one or more artifacts and the other way round.

A part of a configuration model is given in Figure 3-2.

```xml
- <modelConcept name="Software" superconcept="Artifact" />
- <modelConcept name="Common" superconcept="Software" />
- <modelConcept name="Main" superconcept="Software" />
- <modelConcept name="ApplicationModule" superconcept="Software" />
- <modelConcept name="PreFireModule" superconcept="ApplicationModule">
  - <parameter name="Enable" value="{TRUE; FALSE}">
  - <parameter name="realizes" value="PreFire">
  </modelConcept>
- <modelConcept name="PreSetModule" superconcept="ApplicationModule">
  - <parameter name="Distance_Activate_mm" value="[0;20000]">
  - <relation name="realizes" value="PreSet">
  </modelConcept>
- <modelConcept name="ParkingModule" superconcept="ApplicationModule">
  - <parameter name="PARKINGAID_TP_X" value="[0.0;1.0]">
  - <parameter name="PARKINGAID_TP_Y" value="[0.0;1.0]">
  - <relation name="realizes" value="ParkingAssistance">
  </modelConcept>
      ..
- <modelConcept name="Application" superconcept="Feature" />
  - <parameter name="ActivationDistance" value="[0;20000]">
- <modelConcept name="PreSet" superconcept="Application">
  - <parameter name="ActivationDistance" value="[0;20000]">
- <modelConcept name="PreFire" superconcept="Application">
- <modelConcept name="ParkingAssistance" superconcept="Application">
  - <parameter name="Symmetry" value="{Passenger Side; Passenger and Driver Side}">
  </modelConcept>
```

Figure 3-2. Part of a configuration model for the CPS domain

In the XML notation (see Figure 3-2) the concept definitions, their names, superconcepts, and parameters and relations are clearly identifiable. The Symmetry parameter from the example above and some realize relations between the features and software artifacts are given.

5.2   Product Derivation Process

In this section, we explain how software-intensive systems are derived using our methodology. One important difference between the configuration of hardware systems and software-intensive systems is that with software one can realize the product within the derivation process (i.e. compile source code, calibrate the product, etc.). For "traditional" domains, the result of a configuration process is an abstract description of a hardware product. This description is used in manufacturing in order to realize the product. Our model-based product derivation process addresses the entire software development process. Thus, selecting features, architecture and hardware and software artifacts are part of the derivation process just like the realization of the software itself. Therefore two types of activities are involved in model-based product derivation: *configuration* and *realization*.

Configuration activities consist of making decisions about the desired product based on the customer requirements (see Section 5.2.1). A configuration tool is used to ensure a consistent, complete and correct solution. Realization activities produce the software product and other results like a specification document. This is done by selecting artifacts from the asset store, generating new artifacts, compiling new artifacts and compiling the product (see Section 5.2.2).

Configuration and realization do not describe chronological activities. In a typical product derivation process they will rather alternate. For example, once all the features have been selected, which is a configuration activity, the first realization step can take place, i.e. the generation of a specification document.

5.2.1   Configuration

A configuration tool is used to support the configuration of software-intensive systems. One major benefit of the configuration system is that it keeps track of the necessary decisions and computes the currently possible values for these decisions. A further benefit is the computation of values for decisions based on the configuration model. Typically, the user starts by selecting features that represent the functionality of the desired product. In a feature-oriented approach, decisions can be made on a more abstract level and by using customer-understandable terms. As soon as the user has made the first decision, the configuration system starts to compute the impacts of these user decisions. Based on the `is realized by` relation between features and software / hardware artifacts the configuration tool infers the needed artifacts. Thus, in the model-based product derivation process the user makes some decisions and others are inferred by the configuration system.

To illustrate the product derivation process, two screenshots of the ConIPF prototype application are given in Figure 3-3 and Figure 3-4. Usually, a configuration tool is extended with a domain-specific user interface when integrated in an industrial context. For our purposes the tool vendor encoway customized a user interface that suits our methodology. Domain-specific aspects have not been taken into account.

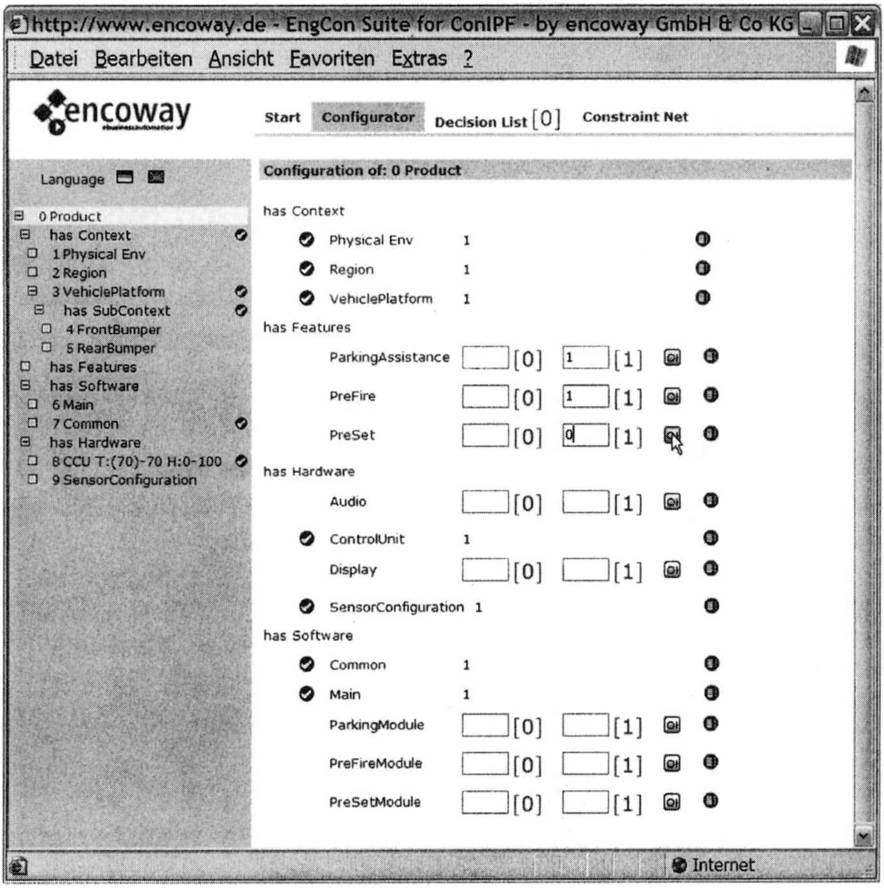

Figure 3-3. Selection of Features (Parking Assistance and PreFire) for a CPS System

On the left side (see Figure 3-3), a tree shows the product derived so far. This tree is structured according to the relations defined in the CAM, e.g. has Feature, has Context and so on. On the right side possible decisions are given. The corresponding part of the solution is highlighted in the tree. The relations defined in the CAM are used to structure the decisions. In this example, three optional features are displayed (Parking Assistance, Pre Set and Pre Fire).

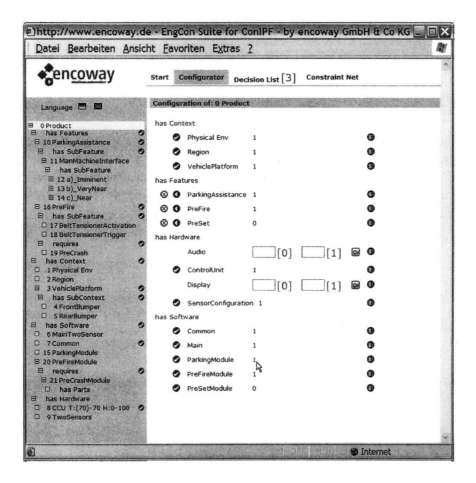

Figure 3-4. Software Artifacts (Parking Module and Pre-Fire Module) inferred by the Configuration Tool

An example scenario from the CPS domain is: the user starts by selecting the features Parking Assistance and Pre Fire and decides not to include Pre Set (see Figure 3-3). Figure 3-4 shows that the values for the selected features have been set and the configuration tool has inferred the corresponding software modules Parking Module (indicated by ID 15) and Pre Fire Module (ID 20) and the feature Man Machine Interface (ID 11) to be part of the CPS system.

5.2.2   Realization

The output of configuration activities is a product description containing all information needed for realizing the product. Because of using the CAM all types of all parts of the product can be identified. For example, the configuration tool EngCon produces exportable XML output that contains a description of the configuration solution. This file contains all information about all assets that have been configured: context, features, and hard- and software artifacts. In Figure 3-5 we show a sample output of the software artifacts that have been configured. The `Product` is the aggregate. It references its parts via the `has Software` relation. Every asset instance has an ID and cardinalities. The `Pre Fire` Module e.g. was not selected and therefore has the cardinality 0 here.

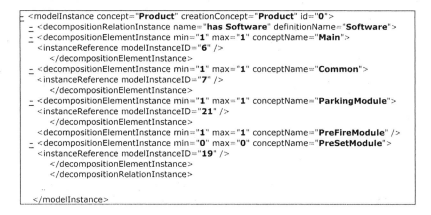

```
<modelInstance concept="Product" creationConcept="Product" id="0">
  <decompositionRelationInstance name="has Software" definitionName="Software">
    <decompositionElementInstance min="1" max="1" conceptName="Main">
      <instanceReference modelInstanceID="6" />
    </decompositionElementInstance>
    <decompositionElementInstance min="1" max="1" conceptName="Common">
      <instanceReference modelInstanceID="7" />
    </decompositionElementInstance>
    <decompositionElementInstance min="1" max="1" conceptName="ParkingModule">
      <instanceReference modelInstanceID="21" />
    </decompositionElementInstance>
    <decompositionElementInstance min="1" max="1" conceptName="PreFireModule" />
    <decompositionElementInstance min="0" max="0" conceptName="PreSetModule">
      <instanceReference modelInstanceID="19" />
    </decompositionElementInstance>
  </decompositionRelationInstance>
  ..
</modelInstance>
```

Figure 3-5. (Part of a) XML configuration output.

The hard- and software artifacts are the assets that have to be assembled to generate the product. To automatically filter the relevant information from the XML structures, a XSL script can be defined which uses the concepts and relations specified by the CAM. Output of that XSL transformation is a list of software artifacts extracted from the configuration result. The source files and their version numbers are extracted from a configuration management system. After compiling the source code the CPS system can be run on the target platform.

In our experiments at Robert Bosch GmbH we used a demonstrator playing sample data that was recorded in a test car. Different asset combinations and different parameter values for selected assets can be tested and compared directly. More detail about the results of our experiments is provided in Section 6. For more details about the product derivation process see (Wolter et al. 2004).

5.3   Adaptation of Artifacts

Although the configuration process always leads to a consistent solution during the process there can be inconsistent partial solutions called *conflicts*. This can happen when e.g. the user selects certain features or components that cannot be combined. A conflict also occurs when an incompatibility is recognized, e.g. when the compilation cannot be executed successfully. Backtracking mechanisms can be applied to take back decisions and their inferences and then try different input for those configuration decisions that led to the conflict. However, it is possible that a conflict cannot be resolved – i.e. no correct solution can be generated for the customer requirements and the given configuration model. In such situations existing assets (and their corresponding description in the configuration model) have to be modified during the derivation process.

In contrast to technical application domains, for configuring software products, even when existing software artifacts are reused, often modifications of those artifacts are needed. The configuration model describes all members of a product family that can be derived using knowledge-based configuration techniques. Thus, the configuration model describes admissible configurations. This can be extended by anticipating future evolution to a certain extent e.g. by modeling planned features (Hein et al. 2001). But eventually there exist unpredicted requirements (like bug fixes) or other situations where evolution planning is not practical.

Evolution during domain engineering is the task of extending the product family, i.e. modeling new variants and versions of components or modifying existing ones. Methods of knowledge acquisition are sufficient for this task. During application engineering, the model is usually fixed for model-based configuration techniques. This means, techniques for dynamically modifying the configuration model have to be taken into account to cope with new functionality during product derivation. Generating solutions that lie outside the modeled solution space is addressed in *innovative configuration* (Günter 1995).

The configuration model can be used for supporting evolution (Krebs et al. 2004a). It reflects all existing components and their dependencies. Thus, the impacts of a component modification or an addition of a component of a specific type can be computed by examining the configuration model (Hotz et al. 2004). A *dependency analysis* has been implemented to compute a graph showing the impacts a modification has. This graph can be used to set up a to-do list containing those assets that have to be modified in order to come to a consistent set of modifications to fulfill the evolution task.

6. EXPERIMENTS AND EXPERIENCES

Both industrial partners, Robert Bosch GmbH and Thales Naval Nederland, have chosen a business unit to apply and validate the methodology. Robert Bosch GmbH has applied the methodology in the area of the CPS domain (that was introduced in Section 4) to develop manufacturable prototypes. Thales Naval Nederland uses the new generation of Combat Management System with the aim of efficiently managing the complexity of the large number of interdependent subsystems. Both partners instantiated the methodology described in this paper, set up configuration models using the modeling facilities defined in Section 5.1 and completed diverse product derivations that led to concrete software products and could be tested on demonstration platforms.

During these experiments, we made the following experiences: the Commonly Applicable Model can be used to model product families for software-intensive systems. The asset types we identified are sufficient to model all aspects needed for automated product derivation support.

During the project subsystems have been identified as helpful for modeling product families. A subsystem can comprise multiple hard- or software assets that are used together leading to a more efficient reuse of frequently used asset compositions (see also Krebs et. al. 2004b).

The methodology can be introduced in different business units and is tailorable for their specific needs. This means it is possible to only use parts of the defined modules (e.g. leave out the evolution of assets) or exchange tools, depending on what kinds of tools or representations are used in that business unit.

The methodology is scalable: while at Bosch we modeled a rather small but complex domain containing about 75 assets and 35 constraints, at Thales the configuration model contains about 700 assets and 250 constraints.

Two major aspects have been observed concerning modeling the product line:

1. Different alternatives to model the same aspect (e.g. modeling a requires relation as a compositional relation or as a constraint) make modeling a non-trivial task. Maintainability is complicated by the fact that the same aspect can be modeled in different ways; e.g. by different persons. For this reason in the final ConIPF methodology concrete modeling guidelines are included (Hotz et al. 2005).

2. It is reasonable to have one domain expert and one modeling expert working together in order to define the configuration model. In our experience this "pair modeling" leads to better modeling results and is more efficient than one of these experts modeling alone.

The model-based product derivation process consisting of configuration and realization activities shows that software products can be reliably assembled. Both industrial partners built multiple products with different feature and artifact selections that could be built and run on demonstration platforms.

A stable product family existing beforehand makes it easier to instantiate our methodology. Setting up a product family, modeling all configurable assets and their interdependencies are time-consuming tasks that need a lot of effort. The dependency analysis we introduced can help in improving the overview of necessary modifications to components in the asset store and the configuration model. This means, its use is not restricted to application engineering but can also be applied for extending the asset store during domain engineering.

7.   RELATED WORK

Well-known approaches for family-based software development are PuLSE (Bayer et al. 1999), KobrA (Atkinson et al. 2000) both developed at the Fraunhofer IESE and FAST (Weiss et al. 1999). The Software Product Line Practice Framework (see SEI) developed at the Software Engineering Institute at the Carnegie Mellon University (SEI) provides a collection of successful approaches rather than a particular methodology.

- **KobrA**: In contrast to our methodology KobrA dose not provide a mapping between features and artifacts. Thus, the knowledge about how to map a specific requirement to software and hardware artifacts is not formalized in this methodology.
- **PuLSE**: PuLSE is a methodology for product family engineering. It supports domain engineering as well as application engineering. Like the ConIPF methodology PuLSE formalizes the mapping between features and artifacts. However, no tool support is given for automatically deriving the needed artifacts for selected features (Bayer et al. 2000). Thus, the ConIPF methodology requires less effort and expert involvement during application engineering than PuLSE does.

Besides the configuration tools EngCon and KONWERK used in the ConIPF project (see Section 3.3) further tools can be used for software product derivation:

- **GEARS**: GEARS is a configuration tool that is specialized on software mass customization (Krueger 2001). It is developed by BigLever Software[3]. The main advantage of the configuration tools used in the ConIPF project over GEARS is that they allow for more complex relations and dependencies between features, context, software artifacts and hardware artifacts.
- **Pure:variants**: Another configuration tool especially for family-based software development is pure:variants provided by pure-systems[4] (Pure-Systems 2003). In contrast to EngCon and KONWERK this tool is restricted to pure software systems, i.e. hardware artifacts and aspects like the context of products cannot be handled.

8. CONCLUSION

In this paper, we have presented a model-based product derivation methodology for application in software-intensive domains. We have introduced the approaches our methodology is based on (i.e. software product families and model-based configuration) and have discussed similarities between model-based configuration and mass customization. We have further shown how product derivation can benefit from a combination of these approaches. The most important aspect of our methodology and also a novelty in this research field is the combination of tool-supported configuration and realization into one process for deriving software products. We also presented first experiences gained from the experiments carried out at our industrial partners from the ConIPF project.

ACKNOWLEDGEMENTS

ConIPF (Configuration in Industrial Product Families) is a three year project that is supported by the EU under the grant IST-2001-34438. Four partners (two industrial and two university partners) are participating in the research work: Robert Bosch GmbH, Thales Naval Nederland, the University of Hamburg and the University of Groningen.

---

[3] www.biglever.com
[4] www.pure-systems.com

REFERENCES

Atkinson, C., Bayer, J., and Muthig, D., 2000, Component-based Product Line Development: the KobrA Approach, *Proceedings of 1ˢᵗ International Software Product Line Conference*. Pittsburg, USA.

Bayer, J., Flege, O., Knauber, P., Laqua, R., Muthig, D., Schmid, K., Widen, T., and Debaud, J.M., 1999, PULSE: a Methodology to Develop Software Product Lines, *Proceedings of the 5th Symposium on Software Reusability*.

Bayer, J., Gacek, C., Muthig, D., and Widen, T., 2000, PuLSE-I: deriving Instances from a Product Line Infrastructure. *7th IEEE International Conference and Workshop on the Engineering of Computer Based Systems*, pp. 237-245. Edinburgh, Scotland.

Bosch, J., Florijn, G., Greefhorst, D., Kuusela, J., Obbink, H., and Pohl, K., 2001, Variability Issues in Software Product Lines. *Proceedings of the Fourth International Workshop on Product Family Engineering (PFE-4)*. Bilbao, Spain.

Davis, S., 1987, *Future Perfect*. Addison-Wesley, Reading, Mass..

Günter, A., 1995. *Wissensbasiertes Konfigurieren*, Infix, St. Augustin.

Günter, A., and Hotz, L., 1999, KONWERK – A Domain Independent Configuration Tool, *Proceedings of Configuration (AAAI 1999) Workshop*, pp. 10-19. Orlando, Florida.

Günter, A., and Kühn, C., 1999, Knowledge-based Configuration - Survey and Future Directions, *Proceedings of XPS-99: Knowledge Based Systems*. Würzburg, Germany.

Günter, A., Hollmann, O., Ranze, K. C. and Wagner, T., 2001, Wissensbasierte Konfiguration von komplexen variantenreichen Produkten in internetbasierten Vertriebsszenarien. *KI*, **15**(1): 33-36.

Hein, A., MacGregor, J., and Thiel, S., 2001, Configuring Software Product Line Features, *Proceedings of ECOOP 2001 - Workshop on Feature Interaction in Composed Systems*, Budapest, Hungary.

Hein, A., and MacGregor, J., 2003, Managing Variability with Configuration Techniques, *Proceedings of the Workshop on Software Variability Management at the ICSE*, Portland, Oregon, USA.

Hollmann, O., Wagner, T., and Günter, A., 2000, EngCon: a flexible domain-independent Configuration Engine. *Proceedings ECAI-Workshop Configuration*, pp. 94-96.

Hotz, L., Günter, A., and Krebs, T., 2003, A knowledge-based Product Derivation Process and some Ideas how to integrate Product Development (position paper*), Proceedings of Software Variability Management Workshop*, pp. 136-140, Groningen, The Netherlands.

Hotz, L., Krebs, T., and Wolter, K. 2004, Dependency Analysis and its Use for Evolution Tasks. L. Hotz and T. Krebs (eds.), *Proceedings of the Workshop on planning and configuration 2004 (PuK-2004)*, Hamburg, Germany.

Hotz, L., Wolter, K., Krebs, T., Deelstra, S., Sinnema, M., Nijhuis, J., and MacGregor, J., 2005, *Configuration in Industrial Product Families – The ConIPF Methodology*. AKA-Verlag, Berlin, to appear.

Kang, K., Cohen, S., Hess, J., Novak, W., and Peterson, S., 1990, *Feature-oriented Domain Analysis (FODA) Feasibility Study*. Technical Report CMU/SEI-90-TR-021, Carnegie Mellon University. Pittsburgh, PA, USA.

Kang, K., Lee, J., and Donohoe, P., 2002, Feature-oriented Product Line Engineering. *IEEE Software* **7/8** : 58–65.

Kiczales, J., Bobrow, D. G., and des Rivieres, J., 1991, *The Art of the Metaobject Protocol*. MIT Press, Cambridge, MA.

Krebs, T., Wolter, K., and Hotz, L., 2004a, Mass Customization for Evolving Product Families, *Proceedings of International Conference on Economic, Technical and Organizational Aspects of Product Configuration Systems*, pp. 79-86, Copenhagen, Denmark.

Krebs, T., Hotz, L., and Wolter, K., 2004b, Pre-Packaged Variability for Product Derivation in Product Lines, *Proceeding of Configuration -- ECAI 2004 Workshop*, pp. 31-33, Valencia, Spain.

Krueger, Charles W., 2001, *Software Mass Customization.*

O, Ying-Lie, 2002, Configuration for mass-customization and e-business. *15th European Conference on Artificial Intelligence, Configuration Workshop.*

Parnas, D.L., 1976, On the Design and Development of Program Families. *IEEE Transactions on Software Engineering*, Vol. SE2, 1, March, pp. 1-9.

Piller, F. T., 2003, What is mass customization. *Mass customization news*, 6(1). Retrieved March 21, 2005, http://www.mass-customization.de/news/news03_01.htm.

Pure-Systems, 2003, *Variant Management with pure::variants.*

Ranze, K., Scholz, T., Wagner, T., Günter, A., Herzog, O., Hollmann, O., Schlieder, C., and Arlt, V., 2000, A structure-based Configuration Tool: Drive Solution Designer DSD. *14. Conf. Innovative Applications of AI.*

SEI (Carnegie Mellon Software Engineering Institute). *A Framework for Software Product Line Practice.* Retrieved March 21, 2005, http://www.sei.cmu.edu/plp/framework.html.

Soininen, T., Tiihonen, J., Männistö, T., and Sulonen, R., 1998, Towards a General Ontology of Configuration, *Artificial Intelligence for Engineering Design, Analysis and Manufacturing,* 1998(12): 357-372, Cambridge University Press, USA.

Thiel, S., Ferber, S., Fischer, T., Hein, A., and Schlick, M., 2001, A Case Study in Applying a Product Line Approach for Car Periphery Supervision Systems, *Proceedings of In-Vehicle Software 2001 (SP-1587)*, pp. 43-55, Detroit, Michigan, USA.

Weiss, D., and Lai, C.T.R., 1999, *Software Product Line Engineering: A Family-based Software Development Process*, Addison Wesley.

Wolter, K., Krebs, T., Hotz, L., and Meijler, T.D., 2004, Knowledge-based Product Derivation Process, *Proceedings of the IFIP 18th World Computer Congress TC12 First International Conference. on AI Applications and Innovations (AIAI2004/WCC2004)*, pp. 323-332, Toulouse, France.

Chapter 4

PRODUCT MODELING ON MULTIPLE ABSTRACTION LEVELS

Kaj A. Jørgensen
Aalborg University, Department of Production

Abstract: Typically, Mass Customization (MC) is most often described in relationship with mass production companies. However, in more and more cases, it is shown what MC means for manufacturing-to-order companies and even for engineer-to-order companies. This paper is initiated by some of the challenges associated with modeling of products and product families in such companies. For some of them, the situation is made extreme by market conditions, which imply long order horizons and many changes of the orders both before and after order acceptance.

With focus on these challenges, an approach is described about modeling of product families on multiple abstraction levels in a way where customer driven product configuration is concentrated on decisions, which are relatively invariant throughout order processing. The approach, which is used here, is based on the theory of general systems and outlined in combination with the abstraction mechanisms classification and composition together with object-oriented analysis and design. A generic model component is presented for enabling representation of models as data models.

Extending a model with more and more details is very typically and is the traditional view derived from the predominant type of modeling tools, e.g. CAD software, where the primary focus is on geometry. It is, however, very important to state that modeling must also be performed on different levels of abstraction. First, the abstraction levels must be identified and, subsequently, each abstraction level must be specified in greater detail.

The modeling approach includes guidelines about how the individual levels can be identified and defined. It is argued that the abstraction levels can be utilized in connection with both analytic and synthetic modeling. Hence, they can be applied to both requirement definition and design by modeling. By this approach, it is also shown how the focus of product configuration must be shifted to identification and definition of attributes instead of modules and components and considerations about the ability to perform the functions, which are required by the customer, are very primary and should be addressed at higher abstraction levels.

Key words: Mass customization, product configuration, product model, product family
                model, abstraction level, information modeling, classification, composition,
                object-oriented analysis and design, module types.

1.   INTRODUCTION

More than one decade ago, Mass Customization (MC) was initiated as a research topic with Davis' publication "From Future Perfect: Mass Customization" (Davis, 1989), presenting how products and services could be realized as a one-of-a-kind manufacture on a large scale. Davis also presented the idea that the customization could be done at various points in the supply chain. In 1993, Pine published a major contribution to the mass customization literature: "Mass Customization: The new Frontier in Business Competition" (Pine, 1993), (Pine et al., 1993), which was an extensive study of how American enterprises during the seventies and eighties had been overrun by the efficient Japanese manufacturers, which could produce at lower costs and higher quality. Since its introduction, MC has called for a change of paradigm in manufacturing and several companies have recognized the need for mass customization. Much effort has been put into identifying, which success factors are critical for an MC implementation and how different types of companies may benefit from it (Lampel/Mintzberg, 1996; Gilmore/Pine, 1997; Sabin/Weigel, 1998; Da Silveira et al., 2001; Berman, 2002).

2.   THE FOUNDATION FOR PRODUCT CONFIGURATION IS PRODUCT MODELING

In order to implement a configuration system, a model describing the product must be defined and implemented, usually in a product configurator. The fact that products must be easily customizable in order to achieve MC has been described comprehensively in the literature. Berman (2002) and Pine (1993) proposed that the use of modular product design combined with postponement of product differentiation would be an enabler to a successful MC implementation. This issue of course also relates to the question of readiness of the value chain.

An often used approach is to describe a series of products building a product family, which is described in one single model. Traditionally, a product family can be viewed as the set end products, which can be formed by combining a predefined set of modules (Faltings/Freuder, 1998; Jørgensen, 2003). The product family model describes which modules are parts of the product family model and how they can be combined. When a product family model is implemented in a configurator, users are allowed to select modules to configure products, and even in some cases the user can select the desired properties of the end product and the configurator selects the corresponding modules (Jørgensen, 2003). Several different methods for defining product models have been constructed during the latest years, each with their own advantages.

A "Procedure for building product models" is described in Hvam (1999) based on Hvam (1994). It is a very practical approach with a seven step procedure, describing how to build a configuration system from process and product analysis to implementation and maintenance. For the product modeling purpose it uses the Product variant master method followed by object-oriented modeling to describe both classification and composition in a product family. The object-oriented approach is also applied by Felfernig et al., (2001), who use the Unified Modeling Language (UML) to describe a product family. This is done by using a UML meta model architecture, which can be automatically translated into an executable logical architecture. In contrast to Hvam (1999) this method focuses more on formulating the object-oriented product structure, rules and constraints most efficiently. The method also focuses on how the customers' functional requirements can be translated into a selection of specific modules in the product family.

Mapping of functional requirements to specific modules is considered in Jiao et al. (1998) and Du et al. (2000), where it is proposed to use a triple-view representation scheme to describe a product family. The three views are the functional, the technical and structural view. The functional view is used to describe, typically the customers, functional requirements and the technical view is used to describe the design parameters in the physical domain. The structural view is used for performing the mapping between the functional and technical view as well as describing the rules of how a product may be configured. The description of this modeling approach is however rather conceptual, and does not easily implement in common configuration tools.

Most of the methods, which exist for modeling configurable products, focus on modeling the solution space of a configuration process. This means that they describe the possible attributes of the products and the product structure as described above. Hence they do typically not focus on information which is not directly used to perform the configuration itself. This information which could include e.g. customer, logistics and manufacturing information are according to Reichwald et al. (2000) similarly important, since a successful implementation of MC must integrate all information flows in the so called "Information Cycle of Mass Customization", which is also presented in Reichwald et al. (2000). Here, the emphasis is put on the importance of managing these flows efficiently, which is most likely to be done by building an integrated information flow. In order to do this, the information must be structured in an appropriate way, which can be done by constructing an information model.

There are of course different strategies on how to construct the most appropriate information models, and they naturally also varies between different companies, markets and products. But even though there is not a single generic strategy from which the optimal information model can be constructed, the importance of this issue must be emphasized because of its importance. Since most of the methods, which are developed for product modeling for MC, have been developed for mass producing companies, these methods are not always easy applicable to other production set-ups. For instance, there is a big difference in what engineer-to-order companies need compared to mass producing companies.

One issue for these companies is the need to concentrate on specification of relatively invariant requirements in the sales process and postpone e.g. the selection of specific components and suppliers as long as possible. This would give the freedom to select the most appropriate components regarding e.g. price as well as make it easier to handle changes late in the process.

3.   MODELS AND MODELLING

Modeling is a very important approach in many design projects where the designed artifact is very complex. Modeling has become even more important because of the fact that computer-based modeling tools have been developed to be more useful and with an increasing number of functionalities. Often, the modeling tools dictate certain modeling methodologies with a number of limitations. However, modeling can be performed in many ways and can have different meanings to designers. The emphasis can be put on many subjects, decisions can be sequenced in many ways and resources can be allocated variously.

Methodologies for system development are often based on concepts derived from General Systems Theory. According to this theory, a *system model* is an intentionally simplified description of a system, fulfilling a certain purpose. Hence, the simplifications imply that some choices are made in order to select the most important properties, components and relationships. Thus, a system model can e.g. be suitable for communication between designers, because with the model, it will be possible to concentrate on the most important aspects of the system.

Models are viewed either as *analytic models*, i.e. models of something existing, often physical or *synthetic models*, i.e. models created as a foundation for construction of something new, which will become physical – an artifact (Jørgensen, 2002). Hence, synthetic models are built purely from ideas, thoughts and imaginations and obtained in some kind of representation. Design by modeling is a development approach, where a synthetic model is designed as an intermediate result and the final result is an implementation of the model in the real world.

One of the most important reasons for synthetic modeling is to be able to manipulate and test the model before the actual physical artifact is built. Modeling makes it possible to ensure that the design is correct and by various presentations of the model at different stages, it is possible to see the consequences of decisions and to reach a good impression of the final result. When synthetic modeling is performed, it is often important to view the model from many different aspects and to represent the model on many different abstraction levels. This is especially necessary at the beginning of the modeling process.

When synthetic modeling is performed, it is often important to view the model from many different aspects and to represent the model on multiple abstraction levels. Abstraction can be used to manage the complexity of modeling. This is especially necessary at the beginning of the modeling process before decisions are made about various details. One way of abstraction is to focus on multiple systems instead of individual systems. Analysis, synthesis and modeling can be performed on individual systems as well as a set of systems as a whole.

4. PRODUCT FAMILY MODELS

A generic model of a product family is termed *product family model*, see figure 4-1 (Jørgensen, 2003). Such a model has a set of open specifications, which have to be decided to determine or configure an individual product in the family. The product family model serves as a foundation for the configuration process and, in order to secure that only legal configurations are selected, the family model should contain restrictions about what is feasible and not feasible. Hence, the product family is the set of possible products, which satisfy the specifications of the product family model. The result of each configuration will be a model of the configured product, *configured product model*. From this model, the physical product can be produced, see figure 4-1. A *product configurator* can be defined as a tool, computer software, which is built on the basis of a product family model and which can support users in the configuration process (Faltings/Freuder, 1998).

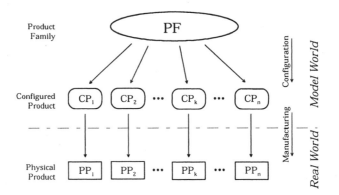

Figure 4-1. The product family model as the foundation for product configuration.

4.1   Product structure – Interfaces

Product configuration in the simplest form is a matter of combining a set of *modules* so that the product model contains information about what modules and components are to be assembled. This *compositional view* declares that a product consists of a number of components, which subsequently can consist of other components, etc. Modules are identified on a level above components from a configuration point of view whereas components usually are identified from a manufacturing point of view. Most often, the number of modules is smaller than the number of related components. Thus, in the *structural model* for configurable products, products consist of modules and modules consist of components. This decomposition into three levels is shown in figure 4-2.

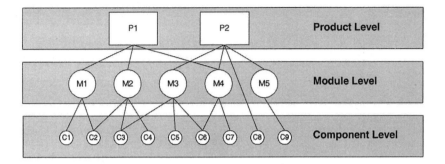

Figure 4-2. Model of the structure with the three levels.

In connection with identification of modules, it is important to analyze how modules interface with each other. Therefore, it is important also to look at the modules functional characteristics and secure that the modular structure is harmonized with the functional division of the product (Andreasen, 2003).

4.2 Properties and attributes

Besides structure, products have *properties*. It is essential for both the customer and the producer to focus on properties of the resulting product. For each configured product, the resulting properties are dependent of the selected components and structure of the product. In the product configuration process, algorithms must be available to estimate the resulting product properties. Some properties are simply the properties of the components, e.g. the color of a car is normally defined as the color of the car body. Other properties are computed from properties of the components. For example, the weight is simply the sum of the component's weight. However, not all resulting properties are so easy to determine. For instance, the resulting performance of a pump is a non-linear function of certain component properties. Much more complicated examples could be mentioned (Männistö et al., 2001).

In the following, the term *attribute* will be used in the models corresponding to properties of physical products. Consequently, when a configuration is performed, the desired properties of the resulting product must be determined by defining values of attributes in the product family model. All relevant attributes of both the resulting product and the available modules must be specified and their optional values to be selected during configuration tasks must also be defined. In relation to this, it is important to notice that the selectable modules and components are sometimes substituted by one or more attributes. For instance, a door can be lockable (attribute) or it can be equipped with a lock (module/component). Therefore, the configuration process can be considered as a mixture of attribute specification and selection of modules, which together can satisfy the required attribute values.

5.   FUNDAMENTAL ISSUES OF INFORMATION MODELING

An important fundamental issue of information modeling is *abstraction mechanisms*, which provide the means for identification and design of invariant components and structures (Smith/Smith, 1977a; Smith/Smith, 1977b; Rosch, 1978; Sowa, 1984). Two abstraction mechanisms are defined here: *composition* and *classification* (Jørgensen, 1998). In essence, composition focuses on the components while classification focuses on attributes. Together, they cover modeling of component *types* as the basis for component *instances* of the information model and they provide the means to set particular focus on the most invariant decisions. A classification process results in a basic hierarchy of types and a composition process results in a basic hierarchy of components. Further, each of the abstraction mechanisms is complemented by two underlying mechanisms: in classification: *generalization* versus *specialization* and in composition: *aggregation* versus *separation*.

Another important issue of information modeling is the *object-oriented paradigm,* which can be adopted in harmony with the abstraction mechanisms. In this paradigm, each model component is regarded as a living organism, which acts and interacts with other components. Thus, object-oriented components are equipped with behavioral attributes, which enable them to respond to requests and, consequently, even if a real world component is non-living, the corresponding model is created as an active component.

Both abstraction mechanisms are used in design tasks, but, as indicated in figure 4-3, classification is used first and composition afterwards. Classification primarily supports the identification of model components and the basic structure at the type level. Based on this, the structural considerations are identified by use of composition.

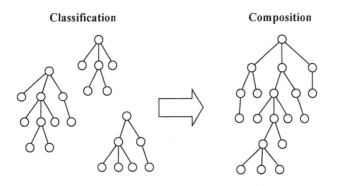

Figure 4-3. Classification and composition hierarchies

6. A GENERIC INFORMATION MODEL COMPONENT

Computer-based models are fundamentally stored in computers as *data objects* and *data structures*, which can be manipulated by applications. Therefore, development of tools for modeling includes both development of a *data model* and a number of *applications* with relationships to the data model (Jazayeri et al., 2000). One of the most important requirements for the data model is that it is non-redundant so that no data value is stored more than once. In order to ensure that this requirement is fulfilled, the model representation has to be considered very carefully based on the meaning of data, the semantics. Therefore, the foundation for a data model is an *information model* (Hammer/McLeod, 1978), created in combination with semantics from the domain, which the design model is addressing.

In order to be able to create all sorts of models and to perform many different modeling processes, a conception of a *generic model component* is introduced. This component is inspired from general systems theory and from object-oriented modeling and can be regarded as a component that can be used for system models in general and for information modeling. Based on this model component, a number of fundamental modeling aspects are described in the following.

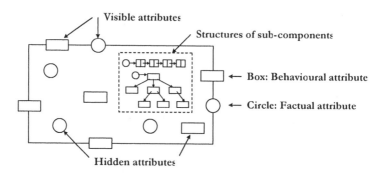

Figure 4-4. Generic model component.

The generic component consists of a set of *attributes* and a *structure of sub-components* (see figure 4-4). Attributes are divided into *factual attributes*, defining the state of the component, and *behavioral attributes*, defining the operations, which the component can carry out. An alternative division of attributes defines some attributes as *visible attributes*, which can be invoked from other components, and some are defined as *hidden attributes*. The structure establishes the relationships between the component itself and the sub-components. All sub-components are regarded the same way, recursively. With this generic component, it is possible to address the following important issues of top-down system modeling: purpose, function (visible behavioral attributes), form (visible factual attributes), internal (hidden) attributes and internal structure.

All structures can be represented by two kinds of relationships in the information model: *references* and *collections*. A reference is a special attribute of which the value holds a direct link to another component. References are very simple and easily understood instruments for solving this because the need for having multiple copies of data values in different components can be eliminated. A collection is also defined as a directed relationship between components. The component, which holds the collection, is the *anchor* component and the components included in the collection are the *member* components. Generally, each collection can consist of zero, one or more member components but, for individual collections, specific constrains can exist. From each anchor component, the members of the body can be accessed as a whole, or individually. Hence, each collection defines an access path from the anchor component to the body components and, to make such access paths as efficient as possible, information structures can be defined.

When a synthetic information model is considered, a foundation for the components must be established by creating *types* of components. Component types are the primary content of information models and it is important to distinguish between modeling on the object level and modeling on the type level. Therefore, a generic component type is also introduced and used in the following. The generic component type is illustrated in figure 4-5 as a box with rounded corners.

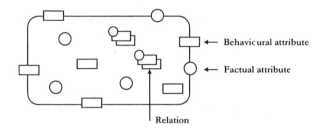

Figure 4-5. Generic component type.

Each component type includes a specification of a set of attributes with *name* and *data type*. The classification abstraction mechanism is primary because, based on attributes, the component types can be classified and organized in a hierarchy, the taxonomy. Identification and specification of structures can also be included in the component types by creating the *relations,* which formulate the *constraints* regarding attributes and combinations of sub-components (see figure 4-5). Simple relations specify *references* and *collections.* The main purpose of references is to create structures, which avoid *redundancy.* When further specifications are added to relations, special component types may be added to the information model. The specifications include semantics like uniqueness, integrity, cardinality, etc.

The component type is a kind of template and, from each type, an indefinite number of components, instances, can be generated (see figure 4-6). The quality of these component types is the key basis to achieve an invariant information model foundation.

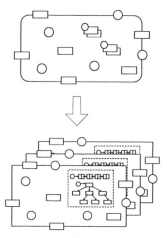

Figure 4-6. Component type is the basis for generating components (instances).

7. FORMULATION OF PRODUCT FAMILY MODELS

As stated above, product family models must be able to construct individual products through a configuration task. Each product model must have sufficient data about attributes and structure in order to manufacture the physical product. Consequently, the basic elements of product family models are the total set of attributes of the possible product models and the set of identified modules, each with their internal attributes and data structures.

The basic units of a product family model are module types. A *module type* is a model of the set of modules, which are interchangeable, perhaps with some restrictions. During configuration, individual modules of each type are selected. The attributes of the module types are selected on the basis of what is important and relevant.

In the following, the contents of product family models are illustrated by use of simple elements of a synthetic language. Furthermore, fractions of a simple example of a computers product family model are added to the illustration. The model content follows the generic model component, which is introduced in the previous section.

Each attribute in a module type is defined by a *name* and usually a *data type* (Boolean, Integer, Float, String, Currency, etc.).

This declarative statement shows the syntax for description of a module type:

name {...}

    Example:
HardDisk {...}

A list of attributes of the module type is described this way:

Attributes
    name : data type;
    name : data type;

    ...

Example:
HardDisk
{
    Attributes
        Name :   String(50);
        StorageCapacity :   Integer;
        AccessTime :   Float;
        Price :   Currency;
}

The available instances of a module type can be listed by a table with a column for each attribute and a row for each module.

Name	StorageCapacity	AccessTime	Price
Maxtor 10K-3	36,7 Gb	4,5 ms	1.375 DKK
Maxtor 10K-4	146,9 Gb	4,4 ms	4.055 DKK
Maxtor 10K-5	300,0 Gb	4,4 ms	8.975 DKK

Alternatively, module data can be extracted from a database. This is also the case for components, which possibly can be modeled by only one module type.

Example:
Component
{
    Attributes
        IdNumber :   Integer;
        Name : String(50);
        CostPrice :   Currency;
}

Some modules can be configured by selecting attribute values. In this
case, each attribute is not defined by a data type but instead by a *domain*
with the possible values. A domain is most often a set of discrete values, an
interval of integer values or a list of named values.

The syntax of an attribute declaration with domain and a possible default
value is:

> *name : {domain} [Default value];*

Example:
```
HardDisk
{
    Attributes
        .....
        PreSet :              {Master, Slave} Default Master;
        OperatingSystem :     {Non, WinXP, Win2000, WinMe}
                              Default WinXP;
        .....
}
```

When module data is specified in form of a table as shown above, the
selection of domain values can be added as columns to the table.

Attributes of a module can be derived from other attributes in the same
module or in other modules. This can be modeled by an *expression* with
standard functions or special functions as a special algorithm. If the name of
a module type is included in such an expression, it means "number of instances of the type".

The syntax of an attribute declaration with domain and a possible default
value is:

> *name : expression [Default value];*

Examples:
```
Computer
{
    Attributes
    Colour =   Case.Colour;
    HardDisks = HardDisk Default 1;
    DiskMemory = Sum(HardDisk.StorageCapacity);
    Weight =   SumWeight : Double
               { ... Specific algorithm ... }
        .....
}
```

Typically for module types, it is possible to add *relations*, which formulate the *constraints* regarding attributes and combinations of sub-modules. In general, there are four different kinds of relations (see figure 4-7).

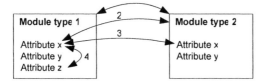

Explanation:

' Constraints between module types

2 Constraints between module types og attributes

3 Constraints between attributes in different module types

4 Internal constraints between attributes in the same modu e type

*Figure 4-7.*Four kinds of relations.

Among other things, relations of category 1 are used to specify product structures. Here, it is described that a product/module (instance of a module type) consists of modules (instances of other module types), which eventually also consist of modules etc. until the component level is reached. In a module type, such a relation expresses the module types for possible submodules. Furthermore, a *multiplicity* is specified in order to form a basic expression about the number of instances that can be included.

Two kinds of multiplicities are available (see figure 4-8).

Multiplicity	Description
OneOf	One and only one instance can be included
AnyOf	One or more instance can be included

Figure 4-8. Domain constraints for attributes that describe if and how the values in the domain shall be chosen.

In addition to the multiplicity, it can be expressed if a sub-module is optional. The syntax for relations describing contents is:

Contents = { [Optional] multiplicity module type, ... };

Example:
```
Cpu
{
    .....
    Contents =
        {OneOf CpuBoard, AnyOf Processor, AnyOf MemoryUnit};
    .....
}

Case
{
    .....
    Contents =
        { OneOf PowerSupply, Optional OneOf PowerCable };
    .....
}
```

All other kinds of relations are formulated by arithmetic expressions. Here, the ordinary arithmetic operators like addition, subtraction, multiplication and division can be used together with standard functions. The following arithmetic relation operators =, >=, <=, >, < and <> can also be used along with the logical operators AND, OR, XOR, NOT, implication ($\Rightarrow$) and bi-implication ($\Leftrightarrow$). If the name of a module type is included in a logical expression, it means "instance of the type".

Examples of relations with arithmetic and logical operators are:
```
Cpu
{
    .....
    Constraints
        GraphicBoard + IoBoard + TvTunerBoard
            <= CpuBoard.NbOfBusSlots;
        Processor <= ProcessorSlots;
        .....
}

Computer
{
    .....
    Constraints
        Monitor <= 2;
        HardDisk + CdDrive + DvdDrive  <=  DiskCable * 2;
```

> OperatingSystem ⇒ HardDisk.OperatingSystem <> Non;
> CdDrive not ⇔ DvdDrive;
>
> .....

}

It must be emphasised that classification and inheritance of attributes can also be included in the development of product family models, especially when many different types of modules are identified.

8. PRODUCT FAMILY MODELING ON MULTIPLE ABSTRACTION LEVELS

When a product or a product family is described, it is often important to see it in a wider context, in relationship with the system (target system) (Jørgensen, 1998), where the product is fulfilling a purpose and performing some desired functions. Hence, this part of a description includes elements from the surrounding environment. Looking solely at a product, a description covers both structure and properties as explained in the previous sections.

To illustrate how multiple higher degrees of abstraction can be identified, a complete description of a product is considered as an analytic model. The generic model component in figure 4-4 can illustrate such a description. Simplification in an analytic modeling approach can be performed along two dimensions: *reduction of structure* and *reduction of properties*. Reduction of structure means identification of upper level systems, which aggregate lover level systems. Ultimately, the internal structure of the system is disregarded completely. Reduction of properties means disregarding subsets of the properties. First internal properties can be ignored so that only the form of the system is considered. Subsequently, a further degree of abstraction would be to focus on the behavioral properties and ignore the factual properties. Such a description is solely aimed at the operations, which can provide transformation of input to output. Hence, at the top level of abstraction, the emphasis should be put on what primarily represents the abilities to perform functions (see figure 4-9).

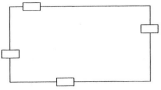

Figure 4-9. Analytic description on top level of abstraction – behavioral attributes.

Applied to synthetic modeling of product families, identification of multiple abstraction levels should be performed by following this general approach in reverse order. In general terms, this will contain considerations according to the following top-down approach: *purpose, function, form (behavioral* as well as *factual attributes)* and *content* (internal *components* and *structure*).

Creation of model components is performed by selection among the component types, which must be available in taxonomies. Which type to select depends on what attributes are required.

Modeling of properties on top level comprises the system's ability to perform functions. This means that the resulting system must be equipped with behavioral properties. Similarly the system model can have corresponding behavioral attributes if, for instance, simulation of the system's behavior is required (see figure 4-10).

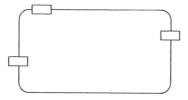

Figure 4-10. Synthetic description on top level of abstraction – behavioral attributes.

In synthetic modeling, however, the functional behavior of the system must be transformed to factual attributes. Hence, each function is characterized and represented by a set of factual attributes, which can be used in connection with specification and test of requirements. In the synthetic model, a component must be created in order to represent the system and the necessary attributes (see figure 4-11).

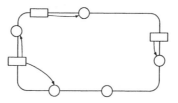

Figure 4-11. Synthetic model on top level of abstraction – behavioral attributes transformed to factual attributes.

With reference to the product family model of computers described in the previous section, a visible attribute 'Wattage' of the module type 'Computer' could represent the function that computers consume electric power. Another visible attribute 'ReadyForUse' could be added to represent that the computer is operable at delivery.

In the next step, internal/hidden attributes and data structures are identified and, finally, the relations of different kinds must be modeled.

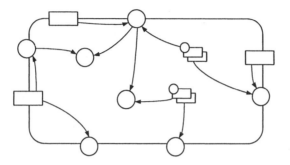

Figure 4-12. Synthetic description of product family model including attributes and relations.

In the computer example, some of the attributes could be defined as hidden attributes if they are only for internal calculations. The constraints are examples of relations. Some of the constraints are only relating attributes to each other and for instance hidden attributes can be related to visible attributes by such constraints. Description of contents is another example of relations and the multiplicity is an example of a further specification of the relation. The remaining constraints either relate attributes to module types or specify relationships between module types.

Figure 4-13 gives an overview of how underlying modules/components of an end-product in a product family can be determined on the basis of decisions regarding attributes.

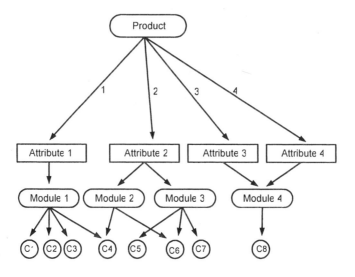

Figure 4-13. Specification of modules directly or indirectly through functionalities.

Attribute 1 corresponds to one module whereas attribute 2 determines two modules. The figure also shows that module 4 is determined by two attributes.

If this idea should be applied to the computer example, all choices about internal modules of the computer must be transformed to attributes. For instance, instead of selecting hard disks directly as sub-modules, at set of attributes must be identified and defined to provide the same possibilities. The attribute 'DiskMemory' is already defined in the 'Computer' module type and through an arithmetic expression it is related to the total storage capacity of the contained disks. With the attribute 'HardDisks' and the table of available disks, it is possible to calculate the obtainable memory values for one disk, two disks, etc. If a value is selected from such a list, the corresponding hard disks are automatically selected and both attributes are updated.

Alternatively, two other attributes could be defined in order to regulate the configuration:

MinDiskMemory : Integer;

LowestDiskPrice : Boolean default true;

This approach of modeling multiple abstraction levels is essential in order to address the challenges, which are identified for engineer-to-order companies and described in section 2. If product configurators are built with greater emphasis put on attributes instead of modules and components, there may be an increased possibility of postponing the decisions about product structure, i.e. specific selection of modules/components. This is particularly important, when the order handling is performed over a long period.

9. CONCLUSION

Implementation of Mass Customization and product configuration in engineer-to-order companies is significantly different compared to mass producing companies. A number of characteristics can be identified. Especially conditions like long order horizons and many changes of order specifications are difficult to manage. In order to develop possible solutions to these problems, a number of theoretical topics about system modeling, product modeling, modeling of product families, information modeling and data modeling must be considered.

From this basis, it is proposed how to perform modeling of product families in a way that multiple levels of abstraction as well as detail can be identified. The result is a synthetic top-down modeling approach with identification and definition of purpose, function, form, content and structure. The aim is to use such models as a foundation for development of product configurator software, which can support long lasting order handling.

As a basis for development of detailed information models, a generic model component is presented. Likewise, a generic component type is introduced as the basis for creation of information models. According to this type, the basic content of product family models is proposed in form of a module type and the use of this module type is illustrated by a number of examples. Additionally by use of the generic component type, some guidelines for identification of multiple abstraction levels of product family models are presented. These guidelines represent possible solutions to the challenges that some engineer-to-order companies are confronted with.

REFERENCES

Andreasen, M. M. (2003): Relations between modularisation and product structuring. In Proceedings of the 6th workshop on Product Structuring – application of product models, MEK-DTU, Denmark, pp. 1-15.

Berman, B. (2002): Should your firm adopt a mass customization strategy? Business Horizons, 45(4):51–60.

Da Silveira, G. / Borenstein, D. / Fogliatto, F. S. (2001): Mass customization: Literature review and research directions. Int. Journal of Production Economics, 72:1–13.

Davis, S. (1989): From future perfect: Mass customizing. Planning Review.

Du, X. / Jiao, J. / Tseng, M. M. (2000): Architecture of product family for mass customization. In Proceedings of the 2000 IEEE International Conference on Management of Innovation and Technology.

Felfernig, A. / Friedrich, G. / Jannach, D. (2001): Conceptual modeling for configuration of mass-customizable products. Artificial Intelligence in Engineering, 15:165–176.

Faltings, B. / Freuder, E. C. (Ed.): Configuration - Getting it right. Special issue of IEEE Intelligent Systems. Vol.13, No. 4, July/August 1998.

Gilmore, J. / Pine, J. (1997): The four faces of mass customization. Harvard Business Review 75 (1).

Hammer, M. / McLeod, D. (1978): The Semantic Data Model: A Modelling Mechanism for Data Base Applications. Proceedings of ACM/SIGMOD International Conference on Management of Data. Austin Texas, pp.144-156, 1978.

Hvam, L. (1994): Anvendelse af produktmodellering, -set ud fra en arbejdsforberedelsessynsvinkel. PhD thesis, Driftteknisk Institut, DTU.

Hvam, L. (1999): A procedure for building product models. Robotics and Computer-Integrated Manufacturing, 15:77–87.

Jazayeri, M. / Ran, A. / van den Linden, F. (2000): Software architecture for product families: Principles and practice. Addison-Wesley, 2000.

Jiao, J. / Tseng, M. M. / Duffy, V. G. / Lin, F. (1998): Product family modeling for mass customization. Computers & Industrial Engineering, 35:495–198.

Jørgensen, K. A. (1998): Information Modelling: foundation, abstraction mechanisms and approach. In: Journal of Intelligent Manufacturing, Vol.9, No.6, 1998. Kluwer Academic Publishers, The Netherlands.

Jørgensen, K. A. (2002): A Selection of System Concepts. Aalborg University, Department of Production, 2002.

Jørgensen, K. A. (2003): Information Models Representing Product Families. Proceedings of 6th Workshop on Product Structuring, 23rd and 24th January 2003, Technical University of Denmark, Dept. of Mechanical Engineering.

Lampel, J. / Mintzberg, H. (1996): Customizing customization. Sloan Management Review, 38:21–30.

Männistö, T. / Soininen, T. / Sulonen, R. (2001): Product Configuration View to Software Product Families. In: Proceedings of Software Configuration Management Workshop (SCM-10). Toronto, 2001.

Pine, B. J. (1993): Mass Customization - The New Frontier in Business Competition. Harvard Business School Press, Boston Massachusetts, 1993.

Pine, J. / Victor, B. / Boyton, A. (1993): Making mass customization work. Harvard Business Review 71 (5), 71(5):108–119.

Reichwald, R. / Piller, F. T. / Möslein, K. (2000): Information as a critical success factor or: Why even a customized shoe not always fits. In Proceedings Administrative Sciences Association of Canada, International Federation of Scholarly Associations of Management 2000 Conference.

Rosch, E. (1978): Principles of Categorisation. In: Cognition and Categorization. Laurence Erlbaum, Hillsdale, Jew Jersey, 1978.

Sabin, D. / Weigel, R. (1998): Product Configuration Frameworks - A survey. In IEEE intelligent systems & their applications, 13(4):42-49, 1998.

Smith, J. M. / Smith, D. C. P. (1977a): Database Abstractions: Aggregation. Communications of the ACM, Vol. 20, No.6. pp. 405-413 New York 1977.

Smith, J. M. / Smith, D. C. P. (1977b): Database Abstractions: Aggregation and Generalization. ACM transactions on Data Base Systems, vol.2, no.2. pp.105-133 New York 1977.

Sowa, J. F. (1984): Conceptual Structures: Information Processing in Mind and Machine. Addison-Wesley, 1984.

Chapter 5

MASS CUSTOMIZATION: BALANCING CUSTOMER DESIRES WITH OPERATIONAL REALITY

Hossam Ismail, Iain Reid, Jenny Poolton, Ivan Arokiam
University of Liverpool, Agility Centre, University of Liverpool Management School (ULMS)

Abstract: Driven by complex social, political, geographic and technological factors, the past decade has seen dramatic changes in the global market environments. Manufacturing companies have been under pressure to meet conflicting goals of efficiency and consumer choice. On one hand customers demand that orders are met faster and at lower cost. On the other hand, they are demanding highly customized products with a wide variety of options. This has led a growing number of economists and scholars to declare that the paradigm of mass production is no longer able to satisfy such demands. As a result new paradigms of agility, responsiveness and mass customization have emerged. Mass customization is the "application of technology and new management methods to provide product variety and customization through flexibility and quick responsiveness at prices comparable to mass-produced products". Mass customization, in itself introduces new demands on firms. These include improved product development processes, flexible manufacturing planning and control systems, and closer supply chain management. Whilst larger organizations by their nature can afford the risk of making mistakes, small to medium enterprises (SME's) are typically more vulnerable, and hence need a structured low risk approach. The second of these shifts is the more relevant to mass customization and often SME's are not able to effectively balance the market needs on one hand and operational efficiencies on the other. In this paper, a method for feature-based mass customization is proposed that translates the voice of the customer into viable integrated product functional requirements, design features, component selection and reuse, and product design modules that are able to provide a better balance between customer requirements and company capabilities at an early stage of product design. The paper demonstrates, via a case study, how the principles of feature-based customization have been adopted by an SME within the context of agility. The paper explores a method for prioritizing the VOC in terms of similarity of functional requirements/features within product families. These consider the factors of design features, modular structures and product component in terms of cost and volume. The 'feature-component matrix' is introduced to represent product families and calculate these similarity coefficients. The goal is to present design and manufacturing engineers with insights into product similarity and feature-based customization. The paper demonstrates, via a case study, how the principles of balancing customer requirements using feature-based customization and how it has been adopted by an SME within the context of manufacturing agility.

Keywords: Voice of the Customer, Product Proliferation, Feature Similarity, New Product Development, Agility

1.   BACKGROUND

1.1   Agility and the Management of Product Portfolios

To remain competitive, manufacturing companies are under tremendous pressures to improve manufacturing efficiency and to provide their customer's with a greater variety of product choice. On one hand customers are demanding that their orders be processed quicker and at lower cost and on the other hand they are demanding highly customized products and variety, therefore, organizations must design products and services that meet or exceed customer expectations. A key question is how manufactures can provide NPD practices within existing product portfolios? This has led a growing number of economists and scholars to declare that the paradigm of mass production is no longer able to satisfy such demands. New paradigms such as portfolio management, mass customization, and agile manufacturing have emerged. This paper presents a review of existing methodologies for assisting in the design of product portfolios and introduces an approach for optimizing the design features prior to the NPD process that can be configured to the needs of manufacturers. The literature and current practice indicate that there have been significant changes made in terms of the manufacturing paradigm shift from traditional manufacturing to a world of agile manufacturing, which is able to respond quickly to customers' demands [1 and 2]. In general, a long product design cycle diminishes the competitiveness of products due to the relatively shortened product lifecycles in the global marketplace. However, the impact on small to medium enterprises (SME's) is not always clear as the resources required to implement such a strategy in an SME often falls beyond what is considered to be acceptable risk.

1.2   Agility and Mass Customization

The ever increasing competition many industries is facing heightened levels of customer desires that demands flexibility, delivery speed and innovation. Agile manufacturing embraces the ability to respond effectively to current market demands, as well as being proactive enough in penetrating new markets. By definition [3, 4] agility is highly dependent on a wide range of operations management capabilities. A large number of publications have emerged in this area. In some it has been introduced as a total integration of the business activities [4] as well as the flexibility of product, people, process and organization. In others it is expressed as agility drivers, capabilities and providers [5]. Agility therefore is a multi-dimensional approach and different subsystems or segments of business may possess different degrees of agility.

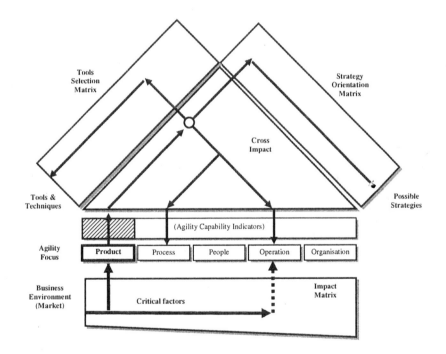

Figure 5-1. Agility Strategic Framework

In view of this the authors have, developed and applied an Agility Strategic Framework (AFS) that integrates the most important aspects of agile manufacturing under one umbrella rather than as individual entities [6]. The approach can be represented as an extended QFD model for agility (see Figure 5-1) in which drivers are the business environmental requirements. Outputs are the tools and strategies that are most suitable for addressing the business environment critical factors. The basic aims of the framework are to provide an approach and a comprehensive set of tools and techniques to support the transitional stages to agility. The driver of the approach is similar to that of TOWs matrix proposed by [7]. The framework's drive towards agility is based on an impact measurement across five business factors these are: product, people, process, operation and organization. It is a generic framework that can be applied to any manufacture company and has been a key tool in implementing a regional agility awareness and implementation program for SME's. The ASF guides the organization through three iterative stages as a route of achieving agility, the three major phases consist of the following:

- **Robustness:** To be able to identify the internal operational vulnerabilities such that risks that have an impact on the business process are reduced
- **Responsiveness:** The ability to convert from a fire-fighting approach enabling the company to respond effectively to customer needs at short notice.
- **Pro-activeness:** Built on freeing up the resources that allows the company to actively seek out new opportunities in existing or new markets.

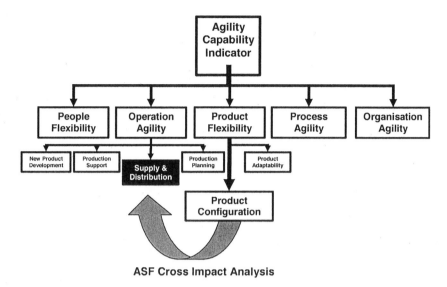

Figure 5-2. Agility Capability Indicators and Product Flexibility

Within the context of the agility strategic framework, mass customization can be viewed as a set of tools for achieving robustness and responsiveness through examining existing product structures and platforms to assess how a degree of modularization and rationalization could be introduced. The aim would be to balance the customer desires and operational reliability. Similarly, it can be used at the proactive stage in the Voice of the Customer in the design of new products and aligning these with current product platforms and company capabilities. Figure 5-2 shows a subset of the agility capability indicators hierarchy. Mass customization impacts on a number off these factors however the focus of this paper will be to describe how product flexibility measures were applied to one SME's as means to assist in the Voice of the Retailer. This leads to the emergence of tools assess the product portfolios of terms of design similarity of product features in terms and component reuse.

2. THE VOICE OF THE CUSTOMER

Approaches to defining product specifications by capturing, analyzing, understanding, and translating the customer requirements, sometimes called the Voice of the Customer (VoC), have received a significant amount of interests in recent years [8]. In order to understand the VoC, Quality Function Deployment (QFD) is one such tool used by engineers, technical development personnel, and quality experts as the ideal tool for capturing the VoC. QFD (see Figure 5-3) is the customer requirements frame to aid the designer's view in defining product specifications. QFD has been heralded as an important part of the product development process. QFD is an investment in people and information. It uses cross-functional teams to determine customer requirements and to translate them into product designs and specifications through highly-structured and well-documented methods.

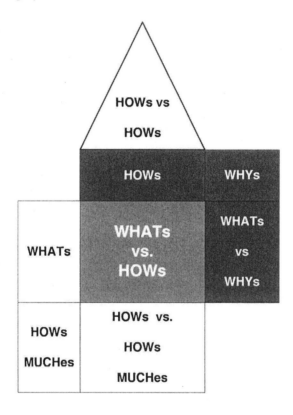

Figure 5-3. The QFD House of Quality

Products designed with QFD may have lower production cost, shorter development time, and higher quality than products developed without QFD. Since its first use, QFD has been widely accepted by a large number of organizations world-wide [14]. Some of its benefits of using QFD include increased efficiency, reduction of cost, shorter lead time, reduction in prelaunch time and after-launch modification and better customer satisfaction [10, 11 and 12]. In spite of the significant number of documented successes with the use with the house of quality (HoQ), there are a number of companies have failed in the process; others have reported mixed experiences [13].

Brown [14] also argued that QFD brings superior product design and the potential for breakthrough innovation. QFD also helps companies discover that innovation, manufacturing and quality can fit comfortably together (Anderson, 1993). While QFD excels in converting customer information to design requirements, it is limited as a means of actually discovering VoC [15]. The VoC in developing product portfolios is investigated in this paper.

2.1   Product Portfolios

Portfolio management is a dynamic decision process, whereby a business will list of active NPD projects and R&D activities are constantly updated and revised. In general, product portfolio planning consists of two stages [16].

1.Product Portfolio Identification: is to capture and understand customer needs effectively and accordingly to transform them into specifications of product offerings (e.g. functional features).

2.Product Portfolio Evaluation and Selection: is to determine an optimal setup or configuration of these planned offerings (e.g. the go/kill decision of an offering) with the objective of achieving best profit performance.

Current researchers and industrial practitioners in this field involve themselves mostly in the economic justification of product portfolio. (e.g. product line design), viz. the latter stage of product portfolio planning. They usually imply the specification of offerings in a product portfolio is given. However, the first issue—how to identify customer needs and generate product portfolio specifications—has received only limited attention. During this phase, many factors are to be considered including any combination of customer needs, corporate objectives, product ideas and related technological capabilities, etc. Usually, product offerings are represented as a list of functional features and target values.

2.1.1   The Product Portfolio Assessment

In order to solve the above problems, the existing product portfolio needs to identity which products within the existing product portfolio are 'Order Qualifiers, Winners and Delighters'. By using the extended Ansoff matrix as a point reference (figure 5-4) there are three main shifts a company carry out to sustain growth:

- Companies traditionally move from sector 1 to sectors 2 and 3 through cost and operational efficiencies and where possible align their existing internal resources and supply chain to meet this new shift.

- A shift from sector 1 to sectors 4, 5 and 6 involves a redesign or modularization of the product to capitalize on new opportunities through customization and product families. A redesign of the product and the supply chain is often required with a shift from cost to flexibility.

- A shift from sector 1 to sectors 7, 8 and 9 are the most risky but offers the company the opportunity to fundamentally redesign both product and supply chain to meet the new product needs.

	New Markets	3	6	9
Existing Market	*New Customers*	2	5	8
	Existing Customers	1	4	7
		Existing Product	**Extended Product**	**New Product**

Figure 5-4. Extended Ansoff Matrix

The second of these shifts is the more relevant to mass customization and often SME's are not able to effectively balance the market needs on one hand and operational efficiencies on the other. In this paper, a method for feature-based mass customization is proposed that translates the voice of the customer into viable integrated product functional requirements, design features, component selection and reuse, and product design modules that are able to provide a better balance between customer requirements and company capabilities at an early stage of product design. The paper demonstrates, via a case study, how the principles of feature-based customization have been adopted by an SME within the context of agility.

Most of the above approaches assume that product development starts from a clean sheet of paper, but in practice, most new products evolve from existing products, i.e. so-called variant design. Historical data and customer data from current products are often considered only implicitly, if not ignored. As a result, product design seldom has the opportunity to take advantage of the wealth of customer requirement information accumulated in existing products. In addition, these methods do not explicitly differentiate the customer preference from the designer's preference of requirement information [17], nor exist in any approach to handling effectively the mapping from the customer domain to the functional domain. Furthermore, new product development in mass customization is facing the challenge of maintaining the continuity of manufacturing and service operations. Therefore, product definition should effectively preserve the strength of product families to obtain significant cost savings in tooling, learning curves, inventory, maintenance, and so on. This demands a structured approach to product definition and to the capturing core information from previous designs as well as existing product and process platforms.

Within the context of a product portfolio, product design and process design are embodied in the respective product and process platforms. Product definition is characterized by the product portfolio representing the target of mass customization (i.e. the 'right' product offerings), which in turn becomes the input to the downstream design activities and is propagated to product and process platforms in a coherent fashion. In this sense, a product portfolio represents the functional specification of product families, i.e. the functional view of product and process platforms [18]. Suh [19] presented a holistic view on product portfolio along the entire spectrum of product development according to the domain framework in axiomatic design which is modified presentation in the case study below. In practice however, many SME's do not have the resources required to employ such formal techniques. Therefore it is more likely that less formal approaches based on subjective intuition and market knowledge will be used.

3.   A METHODOLOGY TO SCHIEVE MASS CUSTOMIZATION

3.1   Feature-based Analysis and Reuse

To minimize internal variety and the associated costs, whilst retaining the external variety that will meet the demands of the customer, modular product structures are developed [20]. The essences of feature-based structures are modules and components that can be selected and combined in order to configure different customized products within a product family. Economies of scale are achieved through the reuse of internal modules and components, rather than finished products of the mass production paradigm. The potential benefits to manufacturing [21] include:

- simplified production planning and scheduling
- lower setup and holding costs
- lower safety stock
- reduction of vendor lead time uncertainty
- order quantity economies

Modular product structures also enable the task of differentiating a product for a specific customer to be postponed until the latest possible point in the supply chain [4]. The postponement of product variety results in risk pooling and consequently reduces overall manufacturing, distribution, and inventory costs [23].

4.   FEATURE SIMILARITY

The ability to identify the most economic functional requirements and maximize their reusability and application within the product portfolio can be achieved through feature similarity. To successfully develop feature-based structures, companies need to measure the similarity between features and the 'reusability' of components in a product family. The measurements should assess the degree of flexibility of a product family, and set boundaries within which new or customized designs will have to function. The measurements should also be used to optimize the use of common components within the product family . One measure that is commonly used is the 'Degree of Commonality Index (DCI)'. This is based on the average number of parent items per distinct component, for a particular product structure [22]. The DCI is defined as:

$$DCI = \frac{\sum_{j=1}^{d} \phi_j}{d} \qquad (1)$$

where: d = the total number of distinct features;
              j = $(1, ..., d)$
              ϕ_j = the number of immediate parents component j has over
a set of end items or product structure level(s)

However, the DCI does not measure the similarity between different functional structures such as different processes and technologies (e.g. electrical circuit and mechanical linkage) structures. Therefore, where a product family contains customizable products with differing functional structures, the DCI cannot be used. The relative commonalities are incomparable.

4.1   The Need for a Measurement of Similarity

To enable companies to develop flexible product features whilst optimizing the reuse of common components and modules, a measurement of the similarity between customizable products within a product family is needed. The objective of reuse is to minimize the total number of parts needed to build the maximum number of customized features. However, this can lead to a conflict. The increasing demands by the customer for greater customization can lead to an increase in the variety of the same product feature (often of low-reuse) used within a product family. This conflict cannot always be resolved. Often a product will contain low-reuse components to differentiate it from other products within the family, or to differentiate it from competitors' products in the market (i.e. external variety). In this situation a compromise should be made. Companies should ensure that low-reuse components are necessary, and can be justified by adding to a product's differentiating attributes. The measurement of similarity between features cannot be based on the reusability of components alone. This is too simplistic and does not account for low-reuse components that add value to a product in terms of differentiation amongst other factors. A more accurate measurement of similarity should also consider factors, such as the cost of the components used to make the product feature, the sales volume of each feature within the family and the to external factors such as suppliers and raw materials.

4.2   Similarity of Product Structures

The 'similarity coefficient' indicate that there have been significant changes made in terms of the manufacturing paradigm shift from traditional manufacturing to a world of agile manufacturing, which is able to respond quickly to customers' demands [1 and 2] This 'similarity coefficient' is based on the product feature structures and reusability of key components and considers the following factors:

- the number of key components in the component group (i.e. the number of components needed to build every feature in the product family) excluding low cost fasteners and standard outsourced components

- the number of distinct components used to build a particular product feature
- the number of other features in the product family using the same components

A similarity coefficient is calculated for each feature with respect to every other feature in the family. The coefficient indicates those products with high-similarity and low-similarity to other features in the product family. Likewise, a feature reuse coefficient is calculated for each component in the group. This should indicate those components with high-reuse and low-reuse.

		Features				
		A1	A2	A3	A4	A5
Product volumes		20	500	300	250	100
Feature Cost		70.00	90.00	110.00	110.00	100.00

		cost	no. used					
	B1	60.00	3	x		x		x
	B2	40.00	3		x	x		
components	B3	10.00	5	x	x	x	x	x
	B4	100.00	2				x	
	B5	30.00	2		x			x
	B6	10.00	1		x			

Figure 5-5. Example Product Feature Structures

In the example features of a product family shown in Figure 5-5 above, product A1 (made from components B1 and B3) is considered to have a high similarity value. All of the components used to make A1 are common to at least three features in the family (i.e. component B1 is common to features A1, A3 and A5). In contrast, feature A4 (made from components B3 and B4) is considered to have a lower similarity value. This is because component B4 is not common to any other feature. In the example above, it can be seen that component B3 has maximum reuse value (i.e. it is common to every feature in the family). In addition to a similarity coefficient that considers product features it is possible to derive similarity coefficients that are based on non-design considerations such as cost and sales volumes as follows:

a. Cost: In practice most product features contain a number of components that are more expensive than others. It is undesirable to have a product features that contain expensive components with a low reuse value unless it is a clear differentiating feature. The effect of component cost can be seen in the example above. Feature A4 (made from components B3 and B4) is considered to have a low similarity value. As already discussed, this is because component B4 is not common to any other feature. In addition, component B4 is the most expensive component in the group.

b. Volume: The sales volume of each feature in a product family also has an impact on similarity. Most product families will contain some features that have low similarity values. This may be for a number of reasons that might include the completeness of a product family, and customer demand. Therefore the cost similarity value of feature A4 is reduced. The effect in terms of sales volume can be also seen in the example above. Feature A2 is made from components B2, B3, B5 and B6. The similarity value of feature A2 is lowered by component B6 (because B6 is not common to any other feature). However, feature A2 has the highest sales volume in the product family. Therefore, the similarity value should reflect this high sales volume, and negate the effects of low similarity based on component reuse.

The 'similarity analysis' can be performed on other non design activities [24] and a number of levels in the product structure and. At the lowest level, this would be for every component in the group. However, for most product families this analysis would be over complex. Therefore it is often necessary to rationalize the component group. Typically, this would usually limit the analysis to those components that collectively contribute to over eighty percent of the product family costs. A 'pareto' analysis can be used for this type of rationalization. At a higher level, an analysis can be performed on any sub-set of the component group. Alternatively, the reuse of particular module or component type can be analyzed.

4.3 The Similarity Matrix

A product family is defined with N distinct components (B1 to Bi) needed to build M finished features (A1 to Aj) within the product family.

N: number of distinct components needed to build the product family
B_i: component (i = 1 → N)
M: number of features in the product family
A_j: product (j = 1 → M)
n_j: the number of distinct components used to build the particular product
m_i: the number of other products in the family, using the same components

The product-component matrix U_{ij} used to represent the product family structure is defined as follows:

$$U_{ij} = 1 \rightarrow B_i \in A_j \tag{2}$$

$$U_{ij} = 0 \rightarrow B_i \notin A_j \tag{3}$$

The number of unique components used to build product Aj is defined as:

$$n_j = \sum_{i=1}^{N} U_{ij} \tag{4}$$

The number of products using component B_i is defined as:

$$m_i = \sum_{j=1}^{M} U_{ij} \tag{5}$$

a.   Feature Similarity Coefficient (R_n):

The '*product structure similarity coefficient*' R_n identifies the similarity of a product with respect to the other products within the product family (i.e. reuse of components). It is only based on the components used in the product structure and defined as follows:

$$R_{nj} = \frac{\sum_{i=1}^{N} U_{ij}(m_i - 1)}{(M - 1)n_j} \tag{6}$$

It is not possible to specify what the minimum level of similarity or variation of similarity within a product family should optimally be. This is dependent on the properties of the product family and the component group analyzed in addition to the minimum requirements for differentiation. The above is also useful in identifying or specifying the core product within a family from which derivatives are generated. Furthermore, the equation only considers distinct components and ignores the quantity of each distinct component in the product. It can be argued that this omission is not critical at this level of analysis as the purpose of the measure is to reduce the number of distinct components. Where the quantity of each component is important is at the rationalization stages where decisions on which components to keep will be influenced by the degree of redesign required.

b. Feature Cost Similarity Coefficient (R_c):
The 'product cost similarity coefficient' R_c identifies the similarity of a product with respect to the other products within the family, based on the costs of the components used in the product structures. Where:

c_i is the cost of component B_i
c_{max} is the maximum cost of all components used in the product family

The 'weighted' cost of component B_i and 'product cost similarity coefficient' R_c, are defined as:

$$w_{ci} = \frac{c_i}{c_{max}} \tag{7}$$

$$R_{cj} = \frac{\sum_{i=1}^{N} U_{ij}(m_i - 1)w_{ci}}{(M-1)\sum_{i=1}^{N} U_{ij}w_{ci}} \tag{8}$$

The 'product cost similarity coefficient' R_c is an improved measurement of similarity because it introduces the factor of cost. The coefficient highlights the effect of costly components that are not reused widely with the product family.

c. Feature Volume Similarity Coefficient (R_v)
The 'feature volume similarity coefficient' R_v identifies the similarity of a feature with respect to the other features within the family, based on the 'sales volumes' of the features used in the product structures. Where:
V_j is the volume of sales for product A_j

The 'weighted' volume for product A_j and the 'weighted' volume of component B_i are defined as:

$$w_{vj} = \frac{V_j}{\sum_{j=1}^{M} V_j} \tag{9}$$

$$w_{vi} = \frac{\sum_{j=1}^{M} U_{ij}w_{vj}}{m_i} \tag{10}$$

The 'product volume similarity coefficient' R_v for a product A_j is defined to be:

$$R_{vj} = \frac{U_{ij} w_{vi}}{n_j} \qquad (11)$$

The *'feature volume similarity coefficient'* R_v introduces the factor of volume to the measure of similarity. Components that are used in products that have a high volume carry more weight. However, as the sales volume increases, R_v has a tendency to distort the measure of similarity. It is possible to achieve high R_v values for products, with a low product structure similarity (i.e. the product is built from 'low-reuse' components). Therefore, R_v should not be used in isolation.

d. Aggregate Feature Similarity Coefficient (R)

The 'aggregate feature similarity coefficient' R combines of the four similarity coefficients: R_n, R_c, R_v and R_t. Each coefficient can be assigned a 'weight' that corresponds to the influence it has on the measure of similarity. Where

w_{rn} is the weight assigned to the 'feature similarity coefficient' R_n
w_{rc} is the weight assigned to the 'feature cost similarity coefficient' R_c
w_{rv} is the weight assigned to the 'product volume similarity coefficient' R_v

The 'product contribution similarity coefficient' R for a product A_j is defined to be:

$$R_j = \frac{R_{nj} w_{rn} + R_{cj} w_{rc} + R_{vj} w_{rv}}{w_{rj} + w_{rc} + w_{rv}} \qquad (12)$$

The weights that are applied to the similarity coefficients R_n, R_c, R_v and R_t. (i.e. w_{rn}, w_{rc}, w_{rv} and w_{rt}) are not fixed and are dependent on the properties of the product family, component types, sales pattern, profit margins etc. It is likely that they will also be unique to each product family. For initial measurements of similarity the 'weights' have been set to unity (i.e. $w_{rn} = w_{rc} = w_{rv} = 1$). The above similarity coefficients are an example of the type of coefficients that can be generated to assess the product family cohesiveness. The measures can be extended to optimize the product family operationally around a group of machines and operations, externally in the selection of suppliers and raw materials, and in optimizing customer requirements in different market segments.

The above similarity coefficients are an example of the type of coefficients that can be generated to assess the product family cohesiveness. Furthermore, the equation only considers distinct components and ignores the quantity of each distinct component in the product. It can be argued that this omission is not critical at this level of analysis as the purpose of the measure is to reduce the number of distinct components. Where the quantity of each component is important is at the rationalization stages where decisions on which components to keep will be influenced by the degree of redesign required. The measures can be extended to optimize the product family operationally around a group of machines and operations, externally in the selection of suppliers and raw materials, and in optimizing customer requirements in different market segments.

5. CASE STUDY

The following section provides an overview of the implementation of the above measures of feature similarity and flexibility to an SME manufacturer. This SME is in the process of implementing agility and mass customization. The company is an OEM and hence directly affected by consumer and retailer demand. The company sells through a small number of powerful retailers who dictate market trends and demand a high level of service. Like many other SME's in a similar situation, this company is driven to be more competitive through the continual introduction of new products and variants on existing products with little planning, resulting in a proliferation of product families and components.

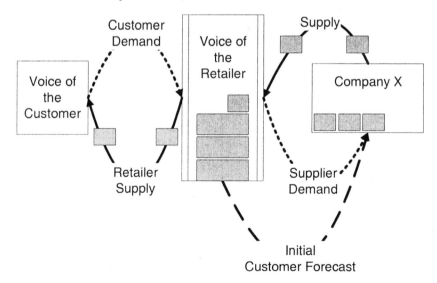

Figure 5-6. Company X's Distribution Channel

5.1   Company Background

Company X is an outdoor toy manufacture that designed and manufactured swings, slides and a variety of play centers. The company trades in a seasonal and highly competitive environment with tight margins and sudden fluctuating customer demands. Over 85% of their sales are generated from only 5 customers who are national retail chains. Each of the customers has a specific design requirement in terms of the type of play activities incorporated in each play centre in addition to a unique color scheme.

Each year, designs are approved individually for each retail customer and a provisional agreement on order quantities is reached. These quantities however can change dramatically with little notice due to uncontrollable factors such as the weather and end-customer purchasing behavior. These retailers have additional requirements such as call off rate and more important cubage. Cubage refers to the dimensional packaging requirements. Demand in the high season outstrips production by a factor of 4. The company has therefore to commit production at an early stage building sufficient stocks to cope with the high season. Consequently, there is a high degree of uncertainty in terms of identifying anticipated sales within each customer product range.

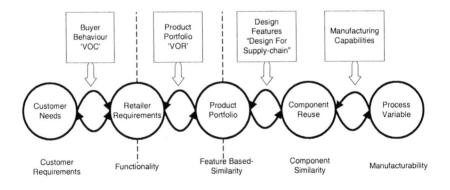

Figure 5-7. Voice of the Retailer's impact on the Product Portfolio
(Modified Model N.P Suh ref 18)

The company has a core production team and a flexible seasonal workforce who are engaged in low skilled activities such as assembly and packing. The company in effect is in a highly vulnerable position with little leverage on suppliers who have specified minimum order quantities and infrequent delivery schedules. This proved to be one of the key problems that limited the company's ability to respond quickly to changes on product.

Early attempts at improvements have resulted in some rationalization of designs and introducing the principle of pack-to-order. However, any benefits from these where overshadowed by inefficiencies resulting from the proliferation of components as new designs are approved with limited reference to existing product ranges. It should be noted once again that the company did not employ a formal technique to establish the optimum modular product structures.

5.2   Similarity Analysis

The analysis below was carried out on the 'Compact' a play centre range that was based on a subset structures from both the swing and slide product families. The purpose was to provide the R&D department with an example of how to analyze product features and to provide them with a means of assessing designs specifically in terms of the supply of raw material. This exercise was particularly aimed at the metal tube components which where proving to be a problem with suppliers as it required a relatively long lead-time for delivery. These where ordered in pre-cut lengths set by the planning department.

Only three of the similarity coefficients were applied (Rn, Rv R) for this analysis. The cost and contribution where excluded as the purpose of the analysis was to rationalize on the supply of raw materials.

Existing Product Family: This product family consists of 11 core design features shown in Table 5-1 The total number of distinct design features in the family, discarding color variations, is 101. These include pre-packed small plastic components and low cost fasteners which for simplicity are treated as single items.

Table 5-1. Feature Contribution of the "Compact" Family product range

						Features					
	F1	F2	F3	F4	F5	F6	F7	F8	F9	F10	F11
N	20	15	16	11	3	8	4	11	4	2	7
Vol	1630	6377	1688	1688	39559	11808	26356	19926	624	1688	608
Cost	£19.78	£13.21	£17.45	£19.31	£31.78	£37.82	£14.88	£11.85	£5.45	£3.74	£8.78

Table 5-2 shows the similarity coefficients across the design features for product structure Rn, sales volume Rv and aggregate R. These results show that feature 6, which has the highest volume and achieves a good rating for across the coefficients. This is due to the fact that the feature has fewer components and these are predominantly shared across the family. At the other end of the scale feature 8 and 9 feature the second highest selling components but have much lower similarity coefficient values with fewer shared components.

Table 5-2. Feature similarity ratings for existing product range

					Features						
	F1	F2	F3	F4	F5	F6	F7	F8	F9	F10	F11
n	20	15	16	11	3	8	4	11	4	2	7
Rn	6.50	9.33	10.0	5.45	3.33	3.75	0.00	0.91	0.00	10.0	0.00
Rv	5.59	7.08	6.41	3.97	36.1	11.1	23.5	17.9	0.56	3.02	0.54
Ra	6.05	8.21	8.21	4.71	19.8	7.43	11.8	9.42	0.28	6.51	0.27

The picture looks far worse for features 1 and 2 if we exclude all non-tube components as shown in table 5-3 where the values of Rn fall to 6.00 and 9.33 respectively representing a very low level of component reuse.

Removal of design features which are process of material characteristics which are different

Table 5-3. Feature similarity ratings for existing product range

					Features						
	F1	F2	F3	F4	F5	F6	F7	F8	F9	F10	F11
n	20	15	16	12	1	8	4	11	4	2	7
Rn	6.00	9.33	10.0	5.00	0.00	3.75	0.00	0.91	0.00	10.0	0.00
Rv	3.83	7.08	6.41	3.77	35.3	11.1	23.5	17.9	0.56	3.02	0.54
Ra	4.91	8.21	8.21	4.38	17.7	7.43	11.8	9.42	0.28	6.51	0.27

The main reason for this is largely due to the fact that this family range uses 56 different tube sizes (diameters, wall thicknesses and cut lengths) and material types as shown in Table 5-4.

Table 5-4. Steel Variations

Material Type	Variations of Outside Diameters	Variations of Wall Thicknesses
E	8	5
G	3	3
R	1	1
S	1	1
T	10	4

Redesigned product features: Further analysis was carried out on the compact product range to identify the effect of two possible changes to the tube components. The first, shown in table 5-5, is where tube cut lengths are ignored and the tube components are grouped by material, diameter and wall thickness. This would represent a case where the company orders tubes in fixed lengths which it cuts to size in house. The total number of tubes in this case would drop from 56 to 30.

Table 5-5. Feature similarity ratings by eliminating pre-cut lengths in both galvanized and non-galvanized tubing

	Features									
	F1	F2	F3	F4	F5	F6	F7	F8	F9	F10
n	6	5	11	7	7	1	6	2	1	3
Rn	20.4	22.2	15.2	17.5	14.3	0.00	11.1	33.3	11.1	11.1
Rv	14.7	20.7	17.8	14.5	19.8	36.4	30.9	25.0	4.66	5.30
Ra	17.6	21.6	16.4	16.0	17.0	18.2	18.2	29.2	7.89	8.21

The second case is where tube of same diameters and materials are standardized utilizing the nearest wall thickness, shown in table 5-6 which reduces the number of tubes further to 23.

Table 5-6. Feature similarity ratings by standardizing tube wall thicknesses

	Features									
	F1	F2	F3	F4	F5	F6	F7	F8	F9	F10
n	5	5	8	6	5	1	4	2	1	3
Rn	28.9	33.3	25.	25.9	26.7	33.3	22.2	50.0	11.1	11.1
Rv	20.7	25.5	26.8	26.3	35.8	57.4	36.9	44.4	4.66	5.30
Ra	24.8	29.4	25.9	26.1	31.2	45.4	29.6	47.2	7.89	8.21

The relatively high similarity values for features 5, 6 and 7 are due to the fact that 2 of their components are common to 6 out of the 8 products and 2 others are common to 4 out of the 8 components.

5.3   Results of the Analysis

The analysis above has highlighted the problem that the company faces in terms of the unnecessarily high number of raw material tubes they use. The problem originates from the wide range of sizes and materials used and this is compounded by the fact that they order these components in pre-cut lengths. With an uncertain demand and long lead-time for delivery the company is exposed to a higher risk of stock out or over stocking. As a result of the similarity analysis the company set up a feasibility study to redesigned internal variety and focus on the following activities:

- Reduced variations of material grades, geometric sizes of outside diameter and wall thickness
- Ordering set lengths, rather than pre-cut lengths

6.   DISCUSSION AND CONCLUSIONS

It is argued that the successful implementation of mass customization starts with the product features that offer customer's choice through optimizing the use of components. With a lack of skill and understanding of how this is carried out SME's often embark on strategy of offering customers more choice without a considered attention to the impact of this on their operations. The monitoring of the implementation process would seem to be critical, SME are unlikely to possess the resources and capacity to achieve everything on their own. The case studies emphasize the importance of external influences on the product change even when people in the company perceive them otherwise.

To cope with customer variety, companies need to balance those customer desires with operational reliability in a flexible manner. At present, SME's often embark on strategy of offering customers more choice without a considered attention to the impact of this on their operations. The above case studies are typical of a few that the authors have observed as part of their involvement with introducing agile practices in SME's. These results are understandable as it is often difficult for SME's, with conflicting measures of performance and a proliferation of tools and techniques, to clearly identify how to proceed. As a simple example, the aim of design for assembly is to optimize product design through simplifying assembly and minimizing the number of components [25]. This is carried out through assessing each component by asking three questions; does the part need to move relative to the rest of the assembly, must the part be made of a different material from the rest of the assembly for fundamental physical reasons, and does the part have to be separated from the assembly for assembly access replacement or repair. If the answer is 'no' to all three then the part is a candidate for removal or integration with adjacent parts. The approach, however, neglects to take into account the possibility that the component is present to provide a differentiation from other products in the family or to extend a core product into a family of products. In effect, mass customized products could individually have more components each but have fewer components collectively within the product family.

A further critical factor that SME's face when embarking on a mass customization program is the number of supplier imposed constraints (e.g. stocking requirements and packing requirements) muffle the true VoC and in supplier to cooperate, as quite often they are competing with larger companies for the supplier's attention. The temptation is therefore to bring processing the outsourced component in-house with the added risk and increased costs it entails. However, a careful assessment of product structures with respect to suppliers can result in identifying those key components that are vulnerable to supplier constraints and increasing their reuse thus providing an incentive for suppliers to cooperate as well as reducing the spread of stock keeping units.

The agility strategic framework was found to be a useful platform for the implementation of mass customization. Through a structured approach, it is possible to identify the critical internal and external factors that inhibit the company's ability to proceed. The classification of those areas affected under product, process, people, operations and organization enables the company to focus on how to mange the implementation process and assess the implication of changes made in any specific areas on other parts of the company. Furthermore the use of the agility capability indicators (ACI) to identify the critical measures of performance avoids the pitfalls of using generic measure of performance that could be misleading.

Furthermore, company X costing structures enable them to clearly quantify the benefits of product feature rationalization or mass customization. For example, the original number of tubing varieties that Company X was manipulating was fifty six. By reducing the variations of material grades, geometric sizes of outside diameter and wall thickness and by ordering set lengths, rather than pre-cut lengths from their suppliers. This was predominantly based on a 'Voice of the Retailer' influence for product differentiation which in this case was high due to the retailers buying power. The considerable benefits of reducing the operational and indirect costs were not recognized.

The paper examines how the reuse of common features in product families can reduce internal variety. Three coefficients that measure the similarity of products within a family are presented. These consider the factors of product structure, cost and volume. The 'product-feature matrix' is introduced to represent product features and calculate the similarity coefficients. The authors have found that these measures are a useful starting point as they set a reasonable baseline for selecting design features when developing new products to existing ones and improvements in similarity coefficients values have a direct affect on product flexibility. In effect the measures were applied to a range of companies including non engineering companies where they were utilized to identify superfluous components and those candidate for integration with others.

REFERENCES

[1]   M.P. Bhandarkar and R. Nagi , STEP-based feature extraction from STEP geometry for agile manufacturing. Comput. Ind. 41 1 (2000), pp. 3–24.

[2]   W.S. Newman, A. Podgurski, R.D. Quinn, F.L. Merat, M.S. Branicky, N.A. Barendt, G.C. Causey, E.L. Haaser, Y. Kim, J. Swaminathan and V.B. Velasco, Jr. , Design lessons for building agile manufacturing systems. IEEE Trans. Robotics Autom. 16 3 (2000), pp. 228–238.

[3]   T.F. Burgess, Making the Leap to Agility: Defining and Achieving Agile Manufacturing through Business Process Redesign and Business Network Redesign, International Journal of Operations and Production Management; Volume 14 No. 11; 1994

[4]   P.T. Kidd , Agile Manufacturing: Forging New Frontiers, Addison-Wesley, MA, 1994

[5]   H. Sharifi, Z. Zhang, "A Methodology for Achieving Agility in Manufacturing Organisations: An introduction International Journal of Production Economics, 1999, pp7-22.

[6]   H. Ismail, S Snowdon, G Vasilakis, I Christian , M Toward "A Strategic Framework for Agility Implementation, IMLF 2002, Adelaide, Australia 2002

[7]   J. Poolton, H. Ismail, "A Marketing Agility Framework for Manufacturing-Based SME's", International Manufacturing Leaders Forum on Global Competitive Manufacturing, Adelaide, Australia, 2005

[8]   A. McKay, A. de Pennington and J. Baxter, Requirements management: a representation scheme for product. Comput-Aided Des 33 7 (2001), pp. 511–520

[9]   Burn, G.R, 1990, "Quality function deployment", Dale, B.G, Plunkett, J.J, Managing Quality, Philip Allan, London.

[10] Sullivan, L.P., 1986, "Quality function deployment", Quality Progress, 19, 6, 39-50.

[11] Hauser, J.R, Clausing, D, 1988, "The house of quality", Harvard Business Review, 66, 3, 63-73.

[12] Zairi, M, Youssef, M.A, 1995, "Quality function deployment: a main pillar for successful total quality management and product development", International Journal of Quality & Reliability Management, 12, 6, 9-23.

[13] A. Griffin and J Hauser, the voice of the customer, technical report, working paper 92-106, Marketing Science Institute, Cambridge, MA 1992).

[14] Brown, N.M, 1991, "Value engineering helps improve products at the design stage", Marketing News, 25, 24, 18.

[15] P.L. Hauge and L.A. Stauffer. ELK. A method for eliciting knowledge from customers, Design and methodology DE-vol . 53 (1993) p. 73–81.

[16] H. Li and S. Azarm, An approach for product line design selection under uncertainty and competition. Trans ASME. J Mech Des 124 3 (2002), pp. 385–392

[17] N.P. Suh, The principle of design. Oxford series on Advanced Manufacturing, 1990.

[18] Jiao and M.M. Tseng, Customizability analysis in design for mass customization. Comput-Aided Des 36 8 (2004), pp. 745–757.

[19] N.P. Suh. Axiomatic design—advances and applications, Oxford University Press, New York (2001).

[20] N. Kohlhase, H. Birkhofer, Development of modular structures: the prerequisite for successful modular products. Journal of Engineering Design, 7(3), pp.279-291, Sep 1996.

[21] C. Sheu, J.G. Wacker, The effects of purchased parts commonality on manufacturing lead time. International Journal of Operations & Production Management, 17(8), pp.725-745, 1997.

[22] E. Feitzinger, H.L. Lee, Mass customization at Hewlett-Packard: the power of postponement. Harvard Business Review, 75(1), pp.116-121, Jan 1997.

[23] M.M. Tseng., M. Lei, C. Su, A collaborative control system for mass customization manufacturing. Annals CIRP, 46(1), pp.373-376, 1997.

[24] J.E. Mooney H. Ismail, S M.M. Shadhidipour "Mass Customisation: A Methodology and Support Tools for Low Risk Implementation in Small and Medium Enterprises, CE2000, Lyon, France 2000

[23] G. Boothroyd, Product Design for Manufacture and Assembly, Marcel Dekker Inc. 1994.

Chapter 6

DESIGN FOR CHANGEOVER (DFC)
Enabling the design of highly flexible, highly responsive manufacturing processes

Michael P. Reik, Richard I. McIntosh, Geraint W. Owen, A.R. Mileham and Steve J. Culley
University of Bath, Innovative Manufacturing Research Centre (IMRC)

Abstract: A rapid changeover capability is central for today's thinking concerning responsive, small batch manufacturing. The customer-driven mass customization paradigm places emphasis on satisfying market demands, particularly in terms of product individualization and ready delivery. Changeover capability is prominent in such a time-based manufacturing environment, where successful companies have to be able to adapt swiftly to market turbulence and at the same time avoid the traditionally high unit costs associated with custom made or small volume products.

Existing tools to improve changeover performance primarily address retrospective improvement, which can be achieved with an emphasis either on refining the activity of those conducting the changeover or changing the hardware that is worked upon. Although there is a choice as to where emphasis might be directed it has been found that retrospective programs are in practice very often led with a strong emphasis on low expenditure and organizational change. Equipment modification opportunities can be significantly undervalued.

Beyond retrospective improvement an excellent changeover capability can also be provided at the outset as an element of overall process equipment capability. The OEM's challenge to build and market changeover-capable equipment is potentially greatly assisted by the availability of a coherent design for changeover (DFC) methodology. Drawing lessons from the development of various DFX methodologies, including design for assembly and design for variety, this chapter discusses the early development of a design for changeover methodology to assist OEMs and other groups responsible for the design and adaptation of process hardware.

Key words: Design for Changeover (DFC), Design for X (DFX), Equipment Design, Responsiveness, Flexibility, Set-up Reduction

1.   INTRODUCTION

Changeover improvement - completing changeover between the manufacture of different products more quickly and to a higher standard - features strongly as a component of modern manufacturing philosophy (Spencer *et al.* 1995; Tu *et al.* 2004). Also referred to as setup reduction or setup reengineering, it is isolated as a core practice when undertaking time-based manufacturing (Koufteros *et al.* 1998). Whereas for traditional mass manufacturing it was sought to minimize production losses by reducing the frequency at which changeovers occurred (Coates 1974), today's more responsive, more flexible models demand that changeover frequency remains high. In turn, if multi-lot production is to be profitable, the duration of the changeover necessarily has to be short.

The rewards of rapid, high quality changeovers are widely described (Schonberger *et al.* 1997). They are advocated as a key instrument to enhance competitiveness, assisting both responsiveness to external market demands (Bicheno 2003) and internal control of factory operations (Suzaki 1987). In either case advantage is primarily gained by making viable the economic manufacture of much smaller batches, ultimately down to lots of just one unit.

Historically changeover performance has been prominent in just-in-time, lean and agile manufacturing models. A rapid changeover capability remains critical in more response-driven mass customization environments, where manufacturers seek to address highly volatile customer demands across a range of customer-tailored products (Pine 1993). In an environment where customers demand niche segmentation crippling disruption to the manufacturing function is likely if downtime attributable to product changeover is not minimized (Abegglen *et al.* 1985).

The authors have determined that the application of design is commonly under-exploited when seeking enhanced changeover performance (McIntosh *et al.* 2001). The rationale of a greater focus on design is elaborated upon within the current chapter, describing as well the important issue of run-up as a component of an overall changeover and the issue of sustaining improved performance. With particular reference to DFMA and other selected DFX methodologies the chapter then goes on to describe the early development of a DFC tool to assist OEM designers. This tool may also be adaptable for retrospective modification of existing process hardware.

2.   MANUFACTURING FLEXIBILITY DEMANDED BY MASS CUSTOMIZATION

Mass customization is a strategy which seeks to exploit market trends for greater product variety and individualization (McCarthy 2004). It is a response to the micro-segmentation of markets and requires that changed practices are introduced throughout, across the whole of the supply chain, manufacturing and marketing processes (Coronado *et al.* 2004). New thinking is needed to the way that businesses are configured so that the necessary responsiveness to market needs can be achieved. For example changed management of market information might be instigated, with far greater transparency, immediacy and customer involvement. Altered information delivery and management systems may require to be installed. Similar considerations apply across all aspects of a successful mass customization operation.

Whereas the need for significant change to the organization is generally recognized, it is also acknowledged that no single good-for-all global model exists. Although certain common elements and 'prescribed formats' are apparent, commonality in implementing a successful mass customization enterprise is lacking (Greenwood *et al.* 1996). Mass customization can be thought of as a still evolving paradigm and yet it is already widely understood that flexibility remains a core requirement - as indeed it is for any responsive multi-product manufacturing environment (Urbani *et al.* 2003). Flexibility is needed across all areas of manufacturing and control, where typically to date there has been an emphasis on personnel and system determinants: continued rigid staff practice and rigid environments imply likely failed implementation, with correspondingly poor response to market demands (Urbani *et al.* 2003).

Although many writers discuss the revised working practices necessary of staff within an organization, it is prudent as well to recognize that enhanced flexibility is a function of the hardware the organization employs. A likely need for better market information management was touched upon above, with possible scope for example for internet utilization by customers, alongside other possible internal use of new electronic systems (McCarthy 2004). This is but one possibility. In the context of flexibility and responsiveness the process hardware itself – the manufacturing equipment used to add value to the product – can be designed to have inherent flexibility. A primary enabler of manufacturing flexibility and responsiveness is manufacturing system changeover.

Together therefore, as elaborated upon below, the performance measured when changing from one product to the next on a given manufacturing line is dependent upon on how people work *and* how amenable the hardware is to being changed over.

2.1   Clarifying organization-led and design-led changeover improvement

Across all facets of manufacturing activity, changeover improvement opportunities can be distinguished between opportunities which are design-led and opportunities which are organization-led (McIntosh *et al.* 1996). Organization-led improvement occurs, for example, when people complete tasks in a more disciplined manner, or when more appropriate tools are used. Organizational improvement can also predominate when the sequence in which tasks are conducted is altered, including arranging for parallel working to occur, or arranging for tasks to be completed in external time. By contrast design-led improvement occurs when there is an emphasis on physically altering manufacturing equipment, thereby, typically, necessarily altering the changeover tasks which previously had to be completed (Rawlinson *et al.* 1996). An amended changeover procedure will necessarily ensue. In some cases, also representing a design change, the product itself can beneficially be altered (Kobe 1992).

2.2   Better changeover capability by original equipment design or by retrospective improvement activity

Discussed above, changeover improvement techniques within industry range from hardware modification through to the control of information and the management, training and motivation of people. The former can be undertaken by original equipment manufacturers (OEMs) or by end users of that equipment. The latter options largely relate to work practices and are often cited as soft or managerial issues. They are issues which OEMs are likely to have little or no impact upon.

Other potential restrictions to attaining an enhanced changeover capability should also be considered. No matter whether OEMs, end users or other agents (for example specialist consultants) are active in seeking improvement there will inevitably be differences in the aptitudes of personnel selected to this task. Tools that personnel might employ to structure or otherwise assist their activity may be deficient. Equally, restrictions may be imposed as to who may participate. Thus for retrospective improvement a business may for example restrict formally trained engineers, or perhaps planning function staff, from becoming involved. Such restrictions have proved to be commonplace during the authors' industrial research and can arise for a variety of often complex reasons (McIntosh *et al.*, 2001). Notably, conventional wisdom dictates that retrospective improvement should be undertaken by empowered workplace teams comprising operators and other direct shop floor personnel whereby continuous, incremental *kaizen* refinement is sought

(Claunch 1996). Additional pressure to adopt organization-biased opportunities can arise when management consultants, training organizations or others leading improvement are under pressure to minimize expenditure (McIntosh *et al.* 2001). Whereas OEMs can only realistically seek to engage design refinement, the uptake of design within conventionally structured SMED programs can be low.

2.3 Shingo's SMED methodology: the primary retrospective changeover improvement tool

The seminal publication on changeover improvement is Shingo's '*A revolution in manufacturing: the SMED system*' (Shingo 1985). The book presents a body of work which forms the basis of innumerable changeover improvement programs in industry, to the extent that the word SMED (Single Minute Exchange of Die) is now commonly encountered, and one which is in itself often taken to mean 'changeover improvement' (Oliver 1989). Wide ranging trade articles, international training agencies and academic attention all indicate that Shingo's staged SMED methodology is far and away the foremost retrospective changeover improvement tool (Herrmann *et al.* 2004).

In spite of its very considerable uptake Shingo's work is not without its critics (McIntosh *et al.* 2000). Design is featured by Shingo as part of the methodology, but this is often communicated as relatively simplistic (although often still valuable) opportunities, such as adding a keyhole slot or substituting a quick release fastener. Further, these opportunities often reside in the methodology's final 'streamlining' phase, and the conceptual purpose of undertaking such changes is not always readily apparent (McIntosh *et al.* 2000).

2.4 Tools available to the OEM

Whereas stronger design guidance might be made available to retrospective improvement practitioners, the need to instruct design for changeover is possibly even more apparent in an OEM context. Although retrospective improvement can be undertaken with either an organizational or a design bias, this option is unavailable to the OEM: original equipment designers can only influence changeover capability by their work prior to equipment installation and commissioning. Their opportunity is that of designing processes with a superior changeover capability. They may possibly also exert influence in some circumstances over product design for changeover.

Many manufacturers have come to appreciate that manufacturing flexibility, via the medium of a competitive changeover capability, is a highly saleable attribute of their equipment. To this end changeover capability is seen to feature in the advertisements of selected OEMs (McIntosh *et al.* 2001). Even with growing recognition of changeover's importance, however, no generic Design for Changeover (DFC) methodology is known. This is an unexpected outcome given both its likely impact and the innumerable other DFX tools which have already been developed. A few companies have made some progress towards a full DFC methodology by refining their own in-house design rules, and some provisional academic work on design rules has also been completed (Van Goubergen *et al.* 2002). Even so, these rules are seen to be considerably lacking in both comprehensiveness and structure in comparison to commercially developed DFX tools.

2.5   A summary perspective

It is clear from the foregoing discussion that generic opportunities for changeover improvement might be classified under the '4Ps' of People, Practice, Process and Products, as shown by figure 6-1 (Reik *et al.* 2005b). These attributes all influence measured changeover performance and are all available to be amended. Thus the motivation of people might be addressed, or better training provided. Or the procedures which are adopted might be revised, for example changing the sequence in which tasks are completed. These are organizationally-biased opportunities, and represent better changeover activity taking place within the confines of largely unchanged hardware. The authors contend the greatest current focus for improvement resides under the categories of People and Practice.

Continuing with this '4P' classification, the products themselves might be revised to enable better changeovers. Equally, physical revision to process hardware is possible. These latter categories are in the realm of design-led improvement.

If the assertion that retrospective programs are organizationally biased is accepted, whereby greater attention is given to attributes of people and practices, then it is reasonable to argue that retrospective improvement agencies and practitioners might be advantaged by a greater awareness of design-focused possibilities. These possibilities could be better structured and communicated than is apparent within today's typical retrospective improvement programs (McIntosh *et al.* 2005).

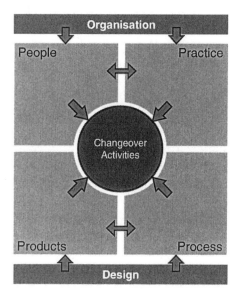

Figure 6-1. The attributes which influence changeover performance

For retrospective changeover improvement all opportunities are potentially available. For the OEM direct opportunities under categorizations of People and Practice are unattainable. Further, within the realm of design-led improvement, minimal guidance is currently available to the OEM.

3. DFC: A REVIEW OF RELATED RESEARCH

The authors' research on Design for Changeover has been informed by the success of other DFX methodologies like Design for Assembly and Design for Manufacture, within which many functional similarities potentially exist. The current section commences by describing a few of these related methodologies and identifying features which might be employed in a changeover context.

Faster and higher quality changeovers will positively impact upon an organization's manufacturing flexibility. Manufacturing flexibility is a highly complex topic (Sethi *et al.* 1990) and includes issues of product updating or replacement (respectively: revised or completely new products). Changeover is usually thought of as a tool to respond faster across an existing and defined product family. Process adaptability to new product families is also investigated, with the intention that the University of Bath DFC work should embody this facility. Thus, following from the work of Slack (Slack 1988), it is sought to have a enhanced changeover capability both in terms of response (changing manufacture between existing products) and range (accommodating the manufacture of new products). Other determinants such as volume flexibility (McIntosh *et al.* 2001) are similarly considered.

Another area of related research is also appraised, extending beyond a review of existing DFX methodologies and manufacturing flexibility. Who can undertake specific types of improvement and what tools are available for their use has already been discussed. The role of design has been investigated, and in doing so the rationale for a DFC methodology established. This rationale is reinforced in the current section where other important issues associated with design are additionally briefly investigated.

3.1   Related design for X Methodologies

The need for 'design for' methodologies was identified as engineers became increasingly aware of a lack of appropriate detailed knowledge in important product life-cycle processes. Design for X methodologies can be seen as tools to analyze designs for their suitability for identified aspects of a product's life cycle. Of these manufacturability and assemblability were among the first to have been considered, since both were highly apparent cost reduction drivers (Huang 1996; Whitney 2004).

Similarly, following the example of DFA and DFM, other DFX methodologies have been proposed which consider alternative life-cycle values, assessing parameters like quality, maintainability, reliability, safety regulations and environmental issues earlier in the design process.

DFX methodologies are tools to evaluate designs, typically at both concept and detail levels. As such, to provide meaningful comparative data, they typically quantify aspects of the design such as cost, quality and regulatory conformity (Reik *et al.* 2004). Thus they not only provide a benchmarking tool for designs but also provide an indication of the possible relative benefits of one design compared to another.

3.1.1   Design for manufacture and assembly (DFMA)

By the simple expedient of the number of hits registered on an internet search engine the researchers determined that the most prominent and widely used DFX methodologies are probably Design for Assembly (DFA) and Design for Manufacture (DFM). DFA provides methods to evaluate ease of assembly, assembly times and costs. DFM helps the designer to enhance manufacturability, and again provides manufacturing cost data for a product and its components. Complex cost models have similarly been developed for different manufacturing processes and their respective process parameters (Boothroyd *et al.* 1994; Swift *et al.* 1997).

(Boothroyd *et al.* 1994) have assessed DFA and DFM separately and explored possible trade-offs between assembly and manufacturing costs. Equipment set-up times are considered in their work and are included in their cost models. However, these data are treated as process related constants in the calculations they undertake. In reality changeover times are strongly dependent on the product range and the manufacturing processes used (McIntosh *et al.* 2001). One objective of DFC is to allow more accurate estimates of the changeover capabilities of process equipment, and the influence that features of the final design will have.

3.1.2 Design for service (DFS)

Design for Assembly (above) in particular comprises likely compatible features with those of DFC, whereby through the use of design it is sought to ease the bringing together of different elements of a larger assembly (be that either a conventionally perceived product or process equipment).

Design for Service (otherwise know as Design for Maintenance) similarly considers how subassemblies can be exchanged as quickly and easily as possible. Depending on the relative likelihood of failure of a specified component or subassembly, greater effort can be made to improve its maintainability. Primarily this is achieved by enhancing ease of both assembly and disassembly, where the additional cost incurred to achieve this objective is justified (Hao *et al.* 2002).

3.1.3 Product Platform Design - Design for Variety, Adaptability and Modularity (DFV)

Research on product platform design seeks to address issues arising from increased customer demand for product variety, shorter time-to market and ever decreasing product life-cycles (Herrmann *et al.* 2004). Design for Variety (DFV) aims to reduce time to market by addressing generational product variation (Martin *et al.* 2002). Indices have been developed for generational variance to help designers reduce the development time of future evolutionary products (Martin *et al.* 2002).

Gu *et al.* (2004) propose a methodology called Adaptable Design which seeks to increase product functionality by increasing the product's adaptability. They discuss that product architecture critically assists process adaptation for a new product. Adaptable Design seeks improvement by segregating the product architecture using platforms, modules and adaptable interfaces.

3.2   Design for Manufacturing Flexibility

Significant current research is being undertaken into changeable and re-configurable manufacturing systems, which are better able to react to each of a full compliment of product characteristics (product individuality). Equally, the manufacturing system needs to be able to respond satisfactorily to variable product volume and mix. Schuh *et al.* (2004) have developed a Design for Changeability methodology which allows manufacturers to assess the degree of process flexibility which should be aspired to. Using a modular approach Schuh *et al.* (2004) distinguish between unstable and stable elements of the production system. Unstable or time variant elements are encapsulated as modules; stable or non-variant elements are encapsulated in platforms.

It is argued that the changeability of a manufacturing system is determined only by a limited number of "change drivers". These change drivers represent variations in product characteristics, capacity requirements, degrees of automation or adaptations which arise due to changes either to statutory requirements or to manufacturing location. The production structure matrix they develop maps change drivers to modules of the production systems and indicates which modules are affected by which change driver. This matrix can then assist in seeking improvements by changing the configuration of the process. The configuration may be changed by integration or separation of production elements or by reduction or (better still) elimination of the influence of a change driver on any selected production element.

Although Schuh *et al.* (2004) offer a useful tool to analyze the correct degree of flexibility for a specific production environment, they do not consider the activities of actually changing the production system from one configuration to a new configuration. A disadvantage of this approach is that it does not allow changeover times to be estimated and thus assess lead times during the OEM design process. Additionally, although Schuh *et al.* (2004) describe the merit of design-led improvements to eliminate influences of change drivers on production elements, no design guidance is provided.

3.3   Further design issues

The rationale for a greater focus on design has been argued so far primarily on the basis that the potential impact of design is under-exploited. Design-led and organization-led opportunities have been differentiated, where design can be applied to process hardware and to the product under manufacture. There are other reasons for a closer focus on design. These reasons further strengthen the case for a DFC methodology for use by OEMs. Described below are other aspects of a changeover which design, if correctly applied, is likely to be particularly adept at addressing. They are aspects of a changeover which the DFC methodology has to embrace.

3.3.1 Changeover quality

Notwithstanding the need for changeovers to be completed as quickly as possible it is also highly desirable for changes to process equipment to be made as accurately as possible, with likely detrimental impact upon both run-up time (below) and subsequent line operation if this does not occur (Sladky 2001). A DFC methodology needs to take the quality to which a changeover is completed into account. Undue concentration on changeover duration can significantly compromise the quality to which a changeover is completed (Smith 1991).

3.3.2 Changeover comprising run-down, set-up and run-up phases

Figure 6-2 shows that an overall changeover can be broken down into three phases - run-down, set-up and run-up. Run-down in particular needs not always be present, but experimental work has proved that run-up is both commonplace and can greatly contribute to lost productivity (Eldrigde *et al.* 2002). The set-up period – the interval when the line is static – is usually expected to occur.

Just as it is important for a DFC methodology to address changeover quality, so too it is important that the methodology is applicable across each of the changeover's distinct phases. Run-up has been found largely to comprise adjustments tasks, with adjustment frequently arising as a result of inconsistent change part setting (Henry 2000).

3.3.3 Sustaining improved performance

In addition it has been shown that design-led improvements undertaken as part of a retrospective improvement program had a pronounced impact on changeover gain sustainability (Culley *et al.* 2003). The key mechanism to sustain changeover gains was design's ability, again if correctly utilized, both to greatly simplify prior changeover tasks and to reduce the number of tasks which needed to be completed. These aims are equally applicable to OEMs.

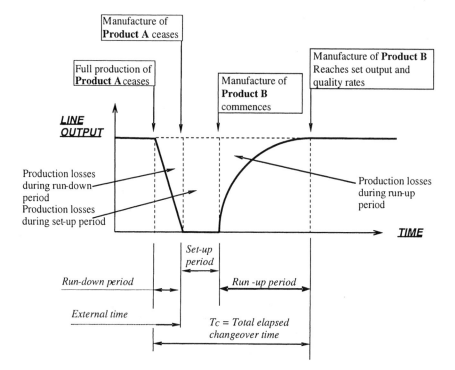

Figure 6-2. Distinguishing phases of a changeover (McIntosh et al., 2001)

4.   ANALYSIS OF CHANGEOVER CAPABILITIES

Analysis of manufacturing equipment in terms of changeover capabilities is fundamental to a successful DFC methodology. This section defines basic entities associated with changeover processes. Relationships between these entities are discussed and the concept of a change driver flow-down is proposed.

4.1   The Changeover Process

The changeover process can be defined as:

A set of activities necessary to correctly set and/or adjust certain elements of manufacturing equipment in order to produce the new product at the desired quality at the desired output rate.

The authors will refer to these elements as change elements, and to the associated activities as changeover activities. Changeover activities in this sense might occur during the three changeover phases, run-down, set-up or run-up, but can also occur before or after, for example as part of the preparation for a changeover.

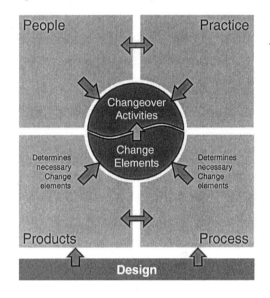

Figure 6-3. Enhanced 4P diagram: Elements of a changeover

The total effort necessary for a changeover is determined by these changeover activities. Required changeover activities are determined by the change elements and their design. This is illustrated by the enhanced 4P diagram in figure 6-3.

If a system's changeover capabilities and general adaptability are to be improved it is necessary to identify the drivers for various changeover activities. These change drivers have a strong influence on the number of change elements required to complete a changeover.

These entities of a generic DFC model, namely *change drivers*, *change elements* and *changeover activities*, are now described in more detail.

4.2   Change Drivers - the Need for Changeover Activity

Manufacturing industry is becoming a more and more customer focused. Ever increasing product variety and shortening product life-cycles forces manufacturers to more strongly consider the impact of these changes on their operations.

The most usual reason for a changeover to occur is to change production facilities from one product to another one, i.e. change is driven by product variation. Product variation can occur on two levels:

- Variation in the product configuration or architecture
- Variation in properties of components, such as form, shape, color, material, process technology for specific features (for example friction or laser welding)

However, there are other drivers which can cause changeover-like activities to occur. One of these drivers can be a change in the output volume. A certain production facility might not be able to manufacture all products at the same output speed. Extra activity might be necessary to set the machine output for a specific product. Also, increased demand for a product by the customer can force the manufacturer to introduce an additional production capability. In a manufacturing line this could for example be achieved by setting up parallel workstations which are manually operated for bottleneck stations.

In addition to changes in product features and output volume, maintenance requirements might mean that certain tools or other parts of a machine need to be exchanged. An example for this has been identified while studying changeover activity on a welding station in a local manufacturing company. Changeovers could be considerably reduced by standardizing the electrode length for all products. However, wear of the electrodes meant that these had to be exchanged during almost any changeover.

Figure 6-4 illustrates the different aspects of change drivers, product mix, output volume and maintenance requirements. Change driven by the product mix can be described by the product structure and product parameters.

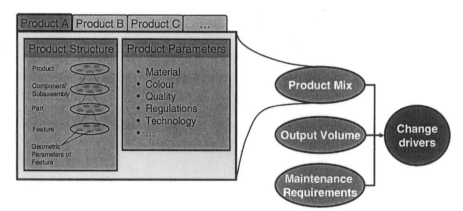

Figure 6-4. Change Drivers: Drivers for changeover activity

Dependent upon the specific change drivers initiating a certain changeover and the relative difference between the involved change drivers before and after the changeover, changeover times often vary considerably. In one case a changeover might only involve changing one machine of a manufacturing line, whereas in another case all line stations have to be changed.

4.3   Change Elements and Changeover Activities

Using the definition of a changeover from section 4.1 it is possible to divide components and modules of process hardware into two types similar to those proposed by Schuh *et al.* (2004) as illustrated in Figure 6-5:

*   Those components which are not affected by any changeover comprise the Equipment Platform.
*   Those modules of manufacturing equipment which undergo changes during a changeover are called change elements.

Change elements (CE) can have two dimensions, representing either physical objects or process parameters. Usually the majority of change elements can be considered as physical objects, like equipment parts (change parts and other parts) or subassemblies. As described by Schuh *et al.* (2004) the concept of a change element can be further expanded to include whole machines or stations. It can even be extended to include lines or sections if considering changeability on different levels within the factory.

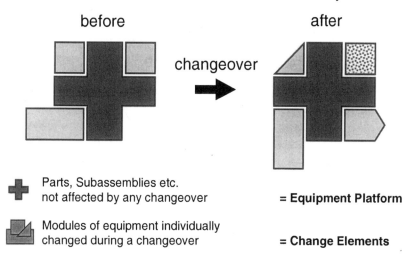

Figure 6-5. Change elements involved in changeover processes

The changeover of a physical object change element can be described by a combination of changeover activities. The different types of changeover activities which can be associated with physical object change elements are described in figure 6-6.

Besides physical objects change elements can also represent other physical entities (process parameters) like levels of electrical voltage, heat flux, hydraulic or pneumatic pressure. A changeover would then be the change from one level of such a unit to another level. Also other activity such as adjustment and the checking or controlling of parameters might be necessary.

In addition to regular disassembly and assembly, the majority of changeover activity comprises setting and adjustment (Shingo 1985). Setting it right-first-time can often not be guaranteed because of insufficient repeatability and accuracy. Reasons for this are variations in the product, its materials or in the process itself. Although there are some cases where product variety cannot be avoided, for example when processing natural food ingredients, process and product designers should generally aim to eliminate this variety.

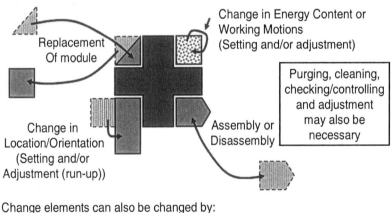

Change elements can also be changed by:
-Energy Content: Temperature, Form of Smart materials, etc.
-Working Motions: Translation/Rotation, Direction, Velocity

Figure 6-6. Changeover activities for physical parts and subassemblies

The following list summarizes the tasks which can be related to the different types of change elements (CEs):

Physical Part

- Disassembly: CE not required anymore or to provide access
- Assembly: CE was not on machine, but is now required
- Positioning: The location and/or orientation of the CE needs to be set
- Energy transfer: Varying energy of CE, for example temperature, velocity, pressure, form for smart materials
- Purging/cleaning: CE needs to be purged or cleaned

Process Parameter

- Setting: CE needs to change in its level and/or value

All Change Elements

- Checking & controlling: Set values need to be checked & controlled
- Adjustment: Adjustment of setting is necessary

The concept of a change element and the association of activities to change elements are fundamental components of the proposed DFC methodology.

4.4   Classifying Change Elements

DFA's central criteria are used to determine whether a part is necessary or unnecessary and therefore whether any possibility for improvement exists. Similarly, criteria can be developed to distinguish between necessary and unnecessary change elements.

Necessary change elements can easily be identified by asking:

"Does this change element have any form of functional contact with the product at any time throughout the entire manufacturing process of the product?"

Each element for which the answer is yes is considered to be a necessary change element (NCE). All other elements are candidates for elimination.

A functional contact between the change element and the product exists if there is an interface between them, that is if there is an interaction between the CE and the product.

A general definition of interfaces (Pahl et al. 1996) categorizes interactions between to interfacing elements as:

- Spatial: Form, shape, location or orientation of the elements define the interface
- Energy flow: Energy is transmitted from one element to the other
- Information flow: Information is transmitted between the participating elements
- Material flow: Material is transmitted between the participating elements

Change elements which do not have any interface with the product throughout the manufacturing process can be of two types. First, the change element is assisting necessary change elements in accommodating to required changes in change drivers. An example would be shims used to locate a machining tool. Second, the change element is only involved in a changeover to provide access or securing for other change elements. Examples for this could be clamping screws or safety covers.

Overall change elements can therefore be categorized as:

A. *Necessary Change Elements (NCE):* CE has interface with product
B. *Indirect necessary CEs (Ind. NCE):*   Assisting NCE in accommodating to required changes in change drivers (for example shims help locating a NCE)
C. *Unnecessary CEs (UCE):*   CE only provides access/securing to other CE

4.5   Relations between Change Drivers and Change Elements –the Change Driver Flow-down

Similar to the production structure matrix proposed by Schuh *et al.* (2004) relations between change drivers and change elements can be described in matrix form (see figure 6-7).

As it has been described in the previous sections a change element type, necessary CE (NCE), indirect necessary CE (Ind. NCE) and Unnecessary CE (UCE), can be associated to every change element depending on a specific change driver.

A. Necessary CE. Interface with product
B. Indirect necessary CE. CE is assisting NCE
C. Unnecessary CE. CE for securing/accessing

change driver			description	Jigs	Binary Decoder	Fastener holding Binary decoder	Electrodes	Centring Bolts for Electrodes	Fastener for Electrodes	Current on Electrode	Computer Disc
Product Features	Product Type		any change (*incl. max size of electrode and material change)	A	B	C	A	B	C	B	B
	Overall Size		smaller	A	B	C	A	B	C	B	B
			larger (*incl. max size of electrode)	A	B	C	A	B	C	B	B
	Wires	Diameter	within tolerances				A			B	
		Material	within tolerances				A			B	
Output Volume/Process Speed			reduce welding operation time (increase size of electrode)				A	B	C	B	B
Maintenance				A			A	B	C	B	
Technology							A	B	C	B	

Figure 6-7. Case study of relations between change drivers and Change Elements

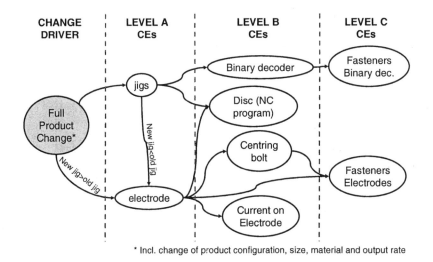

* Incl. change of product configuration, size, material and output rate

Figure 6-8. Sample illustration of a change driver flow-down tree structure

Similar to work proposed by Martin *et al.* (2002) and Whitney (2004) the authors propose a method for the analysis of equipment changeover which decomposes the changes in drivers into the required changes of change elements. A required change in one change element might then be decomposed further into changes of other change elements. These change driver flow-down relationships can best be described by a hierarchical tree structure as shown in figure 6-8. At the top of the hierarchy is a certain change driver or a combination of change drivers. The other levels of the hierarchy are based on the three types of change elements.

The change driver flow-down relationships are based on the information from figure 6-7. Additionally, other information such as assembly and disassembly precedence relationships are considered.

The benefit of the hierarchical representation is the clear guidance it provides as to where to concentrate improvement efforts. Improvement can be undertaken by eliminating the influence of a change driver that is interrupting the change driver flow-down. The higher in the hierarchy this interruption takes place the greater are the potential benefits in respect of changeover. Alternatively, changeover performance can be improved by making changes happen more easily. Again the impact of improvement is potentially far greater if undertaken on higher levels of the change driver flow-down hierarchy (figure 6-8).

5.   EVALUATION OF EQUIPMENT DESIGN

Metrics to evaluate changeover capabilities of existing or proposed equipment designs are an important aspect of a DFC methodology which can be used to quantify the changes in different design proposals.

Earlier sections have described that changeover of manufacturing equipment can be described by the change elements and the changeover activities associated with them.

An approach is proposed where changeovers are analyzed on two levels, one focusing on the change elements, and the other focusing on the changeover activities:

- *The Design Efficiency Analysis*: Equipment design evaluation by comparing the numbers of necessary and all change elements.
- *The Changeover Activities Analysis*: Evaluation of the changeover activities.

It is noted that these DFC analyses concentrate on equipment design. However, a product design's suitability for flexible manufacturing within a certain manufacturing environment can also be benchmarked by using these analyses.

The following sections provide more detailed information on these two types of analyses.

5.1 Design Efficiency Analysis

Based on the identification of necessary change elements, a Design Efficiency Index similar to the design efficiency of DFA methods can be defined (Swift et al. 1997). The DFC Design Efficiency is calculated as the ratio between the number of necessary change elements and the total number of change elements:

(1) $$I_{DE} = \frac{necessary\ CE}{all\ CE} \cdot 100\%$$

5.2 Changeover Activities Analysis

The Design Efficiency Index assists focusing improvement efforts. However, there are trade-offs between the reduction of change elements and the reduction of changeover activities (in some cases it can be beneficial to increase the number of change elements if this significantly reduces the effort involved in changing these elements). Therefore the Changeover Activity Index is proposed which evaluates the effort involved in completing a changeover. In the early stages of a design process typically little detailed information is available about the effort and duration of changeover activities.

Therefore, two strategies have been developed to analyze changeover activities. Strategy A can be used when detailed information of change elements and changeover activities are known to the designer. Strategy B requires less information and is thought to be more relevant for designers during early phases of the design of new manufacturing equipment.

Strategy A
The Changeover Activity Index is calculated as the time ratio of necessary changeover activity to total changeover activity:

(2) $$I_{DE} = \frac{time\ of\ necessary\ Changeover\ Activities}{time\ of\ all\ Changeover\ Activities} \cdot 100\%$$

An activity can only be necessary if it is associated with a necessary change element. Also walking, fetching and adjustment for example are not necessary activities.

Strategy B

In the proactive design of new equipment accurate estimation of change-over times can be difficult. A second strategy has been developed to analyze changeover activities. This approach is based on the scores given to different activities associated to a certain change element. Often it is difficult to make any statement about the effort necessary for changeover tasks. In such a case scores can simply indicate that a certain activity needs to be done. Alternatively, if relative efforts are assumed to be known, scores can indicate the difficulty of changeover activities.

In both cases it can be difficult to estimate the impact of necessary setting and adjustment activities. Therefore, a penalizing mechanism is proposed for change elements with adjustment operations.

It is assumed that all operations associated with the change element in need of adjustment are repeated at least twice in order to manipulate it into its final position. Since checking operations are also necessary it is proposed that change elements in need of adjustment are penalized by multiplying their operation times by a factor of at least three. This can be increased in cases of difficult adjustment, additional checking operations or if scrap is produced.

6.   THE PROPOSED DESIGN FOR CHANGEOVER METHODOLOGY

The aim of the proposed DFC methodology is to provide assistance to OEM designer during the design and development process of manufacturing equipment.

The overall process of such a methodology as proposed by the authors is shown in figure 6-9.

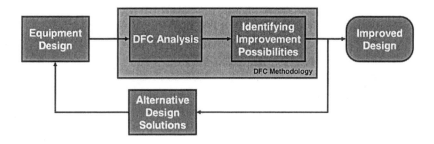

Figure 6-9. Flowchart of the Design for Changeover Process

The 9-step DFC methodology provides guidance for designers from the modeling and evaluation of a changeover process through to identifying improvement possibilities and developing improvement concepts. The methodology concludes with the selection of improvement concepts and evaluation of the improved design. The nine steps are:

1. Identify change drivers
2. Identify change elements and related changeover activities
3. Carry out the DFC design evaluation
4. Identify Relations between Change Drivers and Change Elements
5. Represent relations of Step 4 in a graphical, hierarchical manner
6. Systematic exploration of change element hierarchy for improvement opportunities and creation of design improvement concepts
7. Carry out DFC design evaluations for the proposed improvement concepts
8. Select improvement concepts with the best cost/benefit ratio
9. Carry out the DFC design evaluation for the improved design (Step 3)

These are described in more detail in (Reik *et al.* 2005a).

7. CONCLUSIONS

There is an increasing understanding that leading process changeover performance is essential to modern multi-product manufacturing environments. Changeover performance has been cited as a key element of *just-in-time*, *lean*, *agile* and other manufacturing paradigms. Its role remains prominent in *mass customization*, where particular focus is given to responding to customers needs. A high ensuing level of manufacturing flexibility is required which rapid, high quality changeovers significantly assist.

It has been found that retrospective changeover improvement programs are commonplace within industry. These programs however very often rely disproportionately on aspects of organizational improvement. Similarly design opportunities remain under-utilized by OEM designers. No prior work has been done to provide comprehensive guidance to such personnel in the way, for example, which has been done for design for manufacture or design for assembly. The changeover capability of new process equipment often falls far short of the standard which could be achieved.

Drawing where appropriate upon parallel work in other DFX and related research the authors have set out the basis of a novel design for changeover methodology.

ACKOWLEDGEMENTS

Funding for the Design for Changeover project at the Engineering Innovative Manufacturing Research Centre at the University of Bath has been provided by the Engineering and Physics Research Council (EPRSC).

REFERENCES

Abegglen, J. C. and Stalk, G. (1985). Kaisha: the Japanese corporation. New York, Basic Books.

Bicheno, J. (2003). The New Lean Toolbox - Towards Fast, Flexible Flow. Buckingham, PICSIE Books.

Boothroyd, G., Dewhurst, P. and Knight, W. (1994). Product Design for Manufacture and Assembly. New York, USA, Marcel Dekker.

Claunch, J. W. (1996). Set-Up Time Reduction. Chicago, Irwin Professional Publishing.

Coates, J. B. (1974). "Economics of multiple tool setting in presswork." Sheet Metal Industries 51(2): 73-76.

Coronado, A. E., Lyons, A. C., Kehoe, D. F. and Coleman, J. (2004). "Enabling mass customization: extending build-to-order concepts to supply chains." Production Planning & Control 15(4): 398-411.

Culley, S. J., Owen, G. W., Mileham, A. R. and McIntosh, R. I. (2003). "Sustaining changeover improvement." Proceedings of the Institution of Mechanical Engineers Part B-Journal of Engineering Manufacture 217(10): 1455-1470.

Eldrigde, C., Mileham, A. R., McIntosh, R. I., Culley, S. J., Owen, G. W. and Newnes, L. B. (2002). Rapid Changeovers - The run up problem. ISPE/IFAC Int.Conf. on CAD/CAM, Robotics and Factories of the Future, Porto, INESC Porto.

Greenwood, R. and Hinings, C. R. (1996). "Understanding radical change bringing together the old and new institutionalism." Academy of Management Review 21: 1022-1054.

Gu, P., Hashemian, M. and Nee, A. Y. C. (2004). "Adaptable Design." Annals of the CIRP 53(2).

Hao, J. P., Yu, Y. L. and Xue, Q. (2002). "A maintainability Analysis Visualization System and its Development under the AutoCAD environment." Journal of Materials Processing Technology 129: 277-282.

Henry, J. R. (2000). "Measuring Changeover Success." Journal for Packaging Professionals.

Herrmann, J. W., Cooper, J., Gupta, S. K., Hayes, C. C., Ishii, K., Kazmer, D., Sandborn, P. A. and Wood, W. H. (2004). New Directions in Design for Manufacturing. ASME 2004 Design Engineering Technical Conferences DETC'04, Salt Lake City, USA.

Huang, G. Q. (1996). Design for X - Concurrent Engineering Imperatives. London, Chapman & Hall.

Kobe, G. (1992). "Engineer for right hand steer: how Honda does it." Automotive Industries 172: 34-37.

Koufteros, X., Vonderembse, M. and Doll, W. J. (1998). "Developing Measures of Time-Based Manufacturing." Journal of Operations Management 16(1): 21-41.

Martin, M. V. and Ishii, K. (2002). "Design for Variety: Developing standardised and modularized product platform architectures." Research in Engineering Design 13.

McCarthy, I. P. (2004). "Special Issue editorial: The what, why and how of mass customization." Production Planning & Control 15(4): 347-351.

McIntosh, R. I., Culley, S. J., Gest, G., Mileham, A. R. and Owen, G. W. (1996). "An assessment of the role of design in the improvement of changeover performance." International Journal of Operations and Production Management 16(9): 5-22.

McIntosh, R. I., Culley, S. J., Mileham, A. R. and Owen, G. W. (2000). "A critical evaluation of Shingo's SMED (single minute exchange of dies) methodology." International Journal of Production Research 38(11): 2377-2393.

McIntosh, R. I., Culley, S. J., Mileham, A. R. and Owen, G. W. (2001). Improving Changeover Performance. Oxford, UK, Butterworth-Heinemann.

McIntosh, R. I., Owen, G. W., Culley, S. J. and Mileham, A. R. (2005). "Changeover Improvement: Reinterpreting Shingo's SMED Methodology." (paper accepted for publication by IEEE Transactions on Engineering Management for forthcoming Mass Customisation special issue).

Oliver, M. V. L. (1989). Back to the future - where is JIT going ? 4th Int. Conf. JIT Manuf., Bedford, UK, IFS Ltd.

Pahl, G. and Beitz, W. (1996). Engineering Design - A systematic Approach.

Pine, B. J. (1993). Mass customization: the new frontier in business, Harvard Business Press.

Rawlinson, M. and Wells, P. (1996). Taylorism, lean production and the automobile industry. Beyond modern times. P. Stewart. London, Frank Cass: 189-204.

Reik, M. P., Culley, S. J., Owen, G. W., Mileham, A. R. and McIntosh, R. I. (2004). A Novel Product Performance Driven Categorisation of DFX Methodologies. Advances in Manufacturing Technology, Proceedings of the 2nd International Conference on Manufacturing Research, Sheffield, UK, Sheffield Hallam University.

Reik, M. P., McIntosh, R. I., Owen, G. W., Culley, S. J. and Mileham, A. R. (2005a). "A formal Design for Changeover (DFC) Methodology." Submitted to the Proceedings of the IMECHE - Journal of Engineering Manufacture - Part B.

Reik, M. P., McIntosh, R. I., Owen, G. W., Mileham, A. R. and Culley, S. J. (2005b). The Development of a systematic Design for Changeover Methodology. INTERNATIONAL CONFERENCE ON ENGINEERING DESIGN ICED 05, MELBOURNE.

Schonberger, R. J. and Knod, E. M. (1997). Operations management: customer-focused principles. Boston, Irwin McGraw Hill.

Schuh, G., Harre, J., Gottschalk, S. and Kampker, A. (2004). "Design for Changeability (DFC) - Das richtige Maß an Wandlungsfähigkeit finden." wt Werkstattstechnik 94(04): 100-106.

Sethi, A. K. and Sethi, S. P. (1990). "Flexibility in manufacturing: A survey." Int. J. of Flexible Manufacturing Systems 2: 289-328.

Shingo, S. (1985). A Revolution in Manufacturing: The SMED System. Portland, USA, Productivity Press.

Slack, N. (1988). "Manufacturing systems flexibility - an assessment procedure." Computer-integrated Manufacturing Systems 1(1): 25-31.

Sladky, R. (2001). "Achieving faster, more efficient tube mill changeovers." TPJ-The Tube and Pipe Journal 9(4): 28-31.

Smith, D. A. (1991). Quick die change. Dearborn (Mich.), Society of Manufacturing Engineers.

Spencer, M. S. and Guide, V. D. (1995). "An exploration of the components of JIT: case study and survey results." International Journal of Operations and Production Management 15: 72-83.

Suzaki, K. (1987). The new manufacturing challenge. New York, Free Press.

Swift, K. G. and Booker, J. D. (1997). Process Selection - From Design to Manufacture. Oxford, Butterworth-Heinemann.

Tu, Q., Vonderembse, M. A. and Ragu-Nathan, T. S. (2004). "Manufacturing practices: Antecedents to mass customization." Production Planning and Control 15(4): 373-380.

Urbani, A., Molinar-Tosatti, L., Bosani, R. and Pierpaoli, F. (2003). Flexibility and Reconfigurability for Mass Customization. The Customer Centric Enterprise: Advances in Mass Customization and Personalization Part IV. M. M. Tseng and Piller, F. T. Berlin, Springer.

Van Goubergen, D. and Van Landeghem, H. (2002). Rules for integrating fast changeover capabilities into new equipment design. 11th International Conference on Flexible Manufacturing, Dublin, Elsevier Science Ltd.

Whitney, D. E. (2004). MECHANICAL ASSEMBLIES - Their Design, Manufacture, and Role in Product Development. New York, Oxford University Press.

Chapter 7

PLATFORM PRODUCTS DEVELOPMENT AND SUPPLY CHAIN CONFIGURATION: AN INTEGRATED PERSPECTIVE

Xin Yan Zhang and George Q. Huang
The University of Hong Kong, Department of Industrial and Manufacturing Systems Engineering

Abstract: Platform Products Development (PPD) has been recognized as a formidable approach to effective Mass Customization (MC) for striking the balance between the necessary variety of end-products and the cost to meet the customer requirements in the highly competitive marketplace. Although it is generally acknowledged that it is more effective to consider PPD strategies and design decisions of the associated supply chain simultaneously, little attention and interests have been put on this problem. In this chapter, we conceptualize the problem of integrated configuration of platform products and supply chain and formulate it as a non-linear mathematical problem. Analysis of the model yields two properties, namely *single sourcing* and *absolute replacement*. These two properties form the basis of our solution procedure for the problem. The proposed model and solution procedure are applied to a simple numerical example and the computational results are presented and discussed.

Key words: platform product development; supply chain configuration; mass customization; commonality; modularity.

1. INTRODUCTION

Manufacturing firms often encounter a critical problem of product proliferation. Product proliferation often leads to increased supply chain complexity, unacceptably high production and inventory cost, and long time-to-market. Among the many strategies of mitigating the adverse effects of product proliferation is that of Platform Products Development (PPD), which means developing a set of related products based on a common product platform. PPD is an advanced approach to agile product development (Wheelwright and Clark, 1992; Meyer, 1997; Meyer and Lehnerd, 1997; Robertson and Ulrich, 1998). On the one hand, it is one of the most important means of realizing the MC strategy for creating necessary product variety for competitive success in the marketplace (Salvador et al., 2000). On the other hand, PPD dramatically controls and often reduces not only the cost but also the time-to-market to a competitive level. Leading manufacturers such as Black and Decker and HP have applied some PPD strategies and techniques to rationalize their product lines (Meyer and Lehnerd, 1997).

Despite all the variations, industrial practices in PPD have been based upon a number of strategies. Firstly, the commonality strategy is one of the best known features and also the most important technique of PPD. Through this strategy, the components are standardized and then shared as far as the possible without compromising the variety of the end products entering the market. Secondly, modularity is another essential PPD strategy widely practiced in industries. In this approach, standardized module options are selected and then configured according to specific market and business needs. The modularity strategy implies another PPD technique of multi-functionality. That is, module options are often designed to provide the best proven combination of multiple functions commonly used in a family of products. Apart from the above two basic strategies, postponement and scalability are other widely practiced PPD strategies. In the postponement strategy, the product structures are arranged so that early proliferation of part variety is avoided and variation is allowed and enabled as late as possible in the manufacturing process. The scalability strategy here means "serialization and ranging" of product parameters that have to be changeable. These PPD strategies, which are by no means exhaustive, basically operate on product composition, configuration and characteristics.

Let us define the use of terms *variant module, common module,* and *module option* used in this chapter. It is widely acknowledged that modular product architecture increases flexibility and decreases cycle time in design and manufacturing (Ulrich and Eppinger, 1995). Under the modular product architecture, platform products normally have a fixed number of modules. Each module in turn may have several module options which are somewhat different from each other. The module which has only one module option is a common module and the module which has more than one module options is a variant module. Customization can be achieved through allowing variant modules to choose among a set of given module options. The notion of the variant module here is similar to that of "module type" proposed by Chakravarty and Balakrishnan (2001) and that of "replaceable component set" introduced by Gupta and Krishnan (1999). In this context, customer requirements or optimal PPD decisions can be specified by the module options preferred by the customers or assigned by the manufacturer. In applying platform commonality and modularity strategies, there is a trade-off between providing each market segment of customers with a product exactly satisfying their requirements versus economies of scale among variant end-products achieved by platforming. Platforming, in this chapter, means using a module option with higher performance level instead of that specifically preferred by the customer.

It is widely known that design decisions of products and the associated supply chain are related to each other. However, it is not clear how they interact with each other. This is particularly true when platform products are considered. By applying one or more PPD strategies mentioned above, significant improvements can be made in the associated supply chain of a manufacturing firm. For example, high commonality results in simplified planning and scheduling (Berry et al, 1992), lower setup and holding costs (Collier, 1981, 1982), lower safety stock (Baker, 1985; Dogramaci, 1979), reduction of vendor lead time uncertainty (Benton and Krajewski, 1990) and order quantity economies (Gerchak and Henig, 1989; Gerchak et al, 1988). However, it is not clear how the PPD strategies affect the decisions of Supply Chain Configuration (SCC), and vice versa (Salvador et al, 2002). SCC here means configuring a unique network of supply chain where each node of the network has an assigned option selected from several alternative options for it.

The challenge is therefore how to generate the optimal decisions of PPD and SCC in a simultaneous and integrated manner in order to investigate the mutual impacts between them. The emerging theme, Integrated Configuration of Platform Products and Supply Chain (ICPPSC), has been investigated by some researchers. Gupta and Krishnan (1999) present a decision support methodology for identifying and formalizing the tradeoff between the development costs and benefits of product platforms. Their methodology incorporates the supplier selection decision through one fixed cost of supplier contracting. An optimal set of components are determined firstly and then the set of suppliers are chosen to supply them. In their model, however, they only consider the integration of supplier selection decision into the ICPPSC problem and their solution procedure is somewhat greedy.

Salvador et al. (2002) examine the mutual interactions between product platform strategies (product modularity and variety), production processes and supply sources. Their insights are obtained from empirical case studies. While their findings play important roles in providing general guidance for the decision making process, a quantitative decision is still needed at the tactical decision-making level.

Park (2001) presents a comprehensive model of integrated product platform and global supply chain configuration with experimental simulations. This model has ambitiously incorporated multiple platform strategies and included a large number of supply chain decision variables and parameters along the whole product lifecycle, from the front-end global market segmentation through product design and manufacturing stages, to raw material sourcing and transportation, manufacturing plant location, and end-product distribution. The resulting model is consequently very sophisticated. This leads to difficulties in conducting realistic simulation experiments. Even if meaningful simulations are carried out, it is not easy to independently derive focused findings and insights for decision parameters and variables of primary interest.

In this chapter, we consider a manufacturing firm who is responsible for developing and manufacturing a set of platform products based on the commonality and modularity strategies to satisfy a range of customer requirements. The detailed situation of the manufacturing firm is described as follows: (1) the product platform and the generic product architecture or Generic Bill of Material (GBOM) of the platform products have been established, (2) all the module options of the platform products are produced and supplied by multiple alternative suppliers, (3) the suppliers are independent operators whose decisions are not affected by the manufacturing firm, that is, the unit purchasing prices of module options are fixed by the suppliers. In this condition, the questions faced by the manufacturing firm include: (i) what is the optimal PPD decision, i.e. which kind of end-products to produce; and (ii) what is the optimal SCC decision, such as how much of each module option to order from which supplier at what time interval? In this chapter, these questions are solved through an integrated model.

The rest of this chapter is organized in the following manner. We begin in section 2 with an overview of our research on the ICPPSC problem. Then, we describe the specific ICPPSC problem studied here using an illustrative example in section 3. In section 4, we develop a mathematical decision model to address the problem. The resulting model is a non-linear mathematical programming problem. In our analysis in section 5, we examine the optimality conditions and propose an iterative solution approach based on the properties. Application of the model and the proposed solution procedure to a simple numerical example is discussed in section 6. In section 7, we conclude the chapter by discussion on future work.

2.   INTEGRATED CONFIGURATION OF PLATFORM PRODUCTS AND SUPPLY CHAIN

The research reported in this chapter is actually intended as only a part of our overall research on the ICPPSC problem. This section provides our overall research framework which highlights some background and precondition of the specific research problem studied in this chapter.

Our overall research aims at building a synergy between PPD and SCC decisions mathematically and providing a set of managerial guidelines for optimal PPD and SCC through investigating the mutual impacts between them. At this stage, the precondition of this research is described as follows. The supply chain is composed of a single manufacturer and multiple alternative suppliers. The manufacturer is responsible for designing and manufacturing the platform products with an established GBOM. The suppliers are responsible for supplying raw materials needed by the manufacturer to make the platform products.

As our research scope is comparatively broad, it has been broken down into several scenarios. Three dimensions have been considered, as shown in figure 7-1. The first dimension is the levels of integration or the schemes of coordination of supply chain agents. This dimension falls into two large categories, that is, the agents make their decision in a non-interactive manner or interactively. The non-interactive relationship means that agents make their decision without sharing any information, focusing on their own objectives (e.g. profits) without any consideration of the impacts on their supply chain partners. In this condition, only one agent (e.g. the manufacturer in this research) will be regarded as the rational decision-maker, whereas suppliers' decisions (e.g. unit purchasing prices of module options) are independent of the manufacturer's decisions. In other words, the manufacturer makes its own PPD and SCC decisions with no regard to its impact on the suppliers.

The interactive relationship, on the contrary, means that there exists some kind of information sharing among agents. In this condition, both the manufacturer and the suppliers are considered as rational decision-makers. Specifically, the interactive relationship can be classified into several types according to the levels of agent interaction. They include non-cooperative, cooperative, and fully integrated, with increasing levels of supply chain integration. In a non-cooperative supply chain, agents may share information but they only optimize their own objectives. Sometimes, this process is dominated by one agent (the leader) only and other agents (followers) follow the leader. In a cooperative supply chain, agents not only share information but also objectives while they are autonomous decision makers. They aim to optimize the total objective of the entire supply chain while negotiating the fair sharing of the extra values. In a fully integrated supply chain, there is only one decision model and a single objective for the entire supply chain. Agents no longer have their autonomous decision models or individual objectives. Some agents may benefit / loose more than others and they do not fairly share the benefits due to the full integration. Which level of integration or coordination scheme the supply chain agents should agree upon and adopt is itself a tough area for research.

The second dimension, as indicated by the vertical arrow pointing downwards in figure 7-1, is about PPD strategies. Along the vertical arrow, the commonality decreases while the customizability increases. PPD strategies (e.g. commonality, modularity, postponement, and scalability) included in this dimension have been discussed in the introductory section of this chapter.

The third dimension, as indicated by the horizontal arrow pointing rightwards in figure 7-1, is about SCC (supply chain configuration) decisions. Typical SCC decisions include, but are not limited to, supplier selection, inventory allocation, ordering policy, operation selection, service time, and quantity discount, etc. Along the horizontal arrow, increasing SCC decisions would be included in this research.

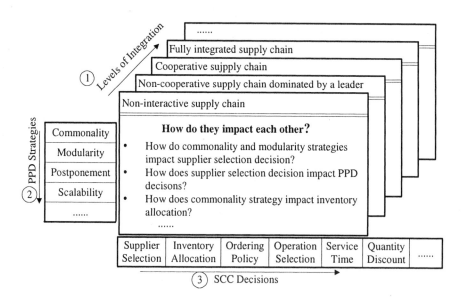

Figure 7-1. Typical scenarios of Integrated Configuration of Platform Products and Supply Chain (ICPPSC)

One of the main contributions made in this research is a common and consistent framework for considering these different scenarios. Firstly, relationships between chosen PPD strategies and SCC decisions must be considered and research problems must be specified. Then, mathematical model(s) will be established for each decision-maker. After that, a solution algorithm that is able to produce effective optimal solution should be developed based on the mathematical model(s). The resulting mathematical model(s) and solution algorithm are then used to conduct series of simulation experiments and sensitivity analyses. Finally experimental results should be analyzed to derive some managerial implications. This will allow us to draw comparative analyses and address the question how supply chain coordination schemes affect the PPD and SCC decisions.

Following the above ICPPSC map, we have so far finished research work on two selected scenarios. One is the scenario of a non-interactive supply chain (Huang et al., 2005a). In this scenario, (1) the agent relationship is non-interactive, (2) only the commonality strategy is chosen for consideration, and (3) chosen SCC decisions include supplier selection, inventory allocation, operation selection, and service time. The other scenario is a non-cooperative supply chain dominated by the manufacturer as the leader (Huang et al 2005b). In this scenario, (1) the agent relationship is non-cooperative, (2) the commonality and modularity strategies are chosen for consideration, and (3) chosen SCC decisions include supplier selection, inventory allocation, ordering policy, and quantity discount.

In this chapter, the scenario of another non-interactive supply chain is investigated. In this scenario, (1) the agent relationship is non-interactive, (2) the commonality and modularity strategies are chosen for consideration, and (3) chosen SCC decisions include supplier selection, inventory allocation, and ordering policy.

3. PROBLEM ILLUSTRATION

In this section, we describe the specific ICPPSC problem of a manufacturing firm studied in this chapter using an illustrative supply chain. As can be seen schematically in figure 7-2, we consider a Manufacturer who plans to develop a set of platform products (PP), which are composed of a set of known modules that are designed and connected by the Manufacturer and produced and supplied by the suppliers, in order to satisfy customer requirements of market segments. Although figure 7-2 is schematic, for example in the context of computer design, one can think of m_1 as being CPU, m_2 as a hard disk module, and m_3 as the motherboard. Among these modules, m_3 is a common module as it is common to each variant in the platform products. The two variant modules, m_1 and m_2, each of them has three module options. Here we introduce the first important assumption of this chapter.

Assumption 1: Module options of each variant module can be arranged in an increasing order of a certain performance characteristic, and the development cost and purchasing price of the higher-ordered module option are always higher than the lower-ordered module option.

For the platform products in figure 7-2, there are 6 alternative suppliers. For simplicity, we assume each supplier can only produce one kind of variant module and its capability or flexibility is measured by the number of module options it can supply. Specifically, we consider two kinds of suppliers according to their flexibility levels: high and low. The supplier with high flexibility can supply all the module options of a variant module while the supplier with low flexibility can only supply either the lower half or the higher half of module options of a variant module. In the example shown in figure 7-2, S_3 and S_6 are suppliers with high flexibility, while S_1, S_2, S_4 and S_5 are suppliers with low flexibility. In this chapter, we assume that:

Assumption 2: The capacity of each supplier is unlimited.

The module options purchased from the suppliers are assembled by the Manufacturer into assemblies and then end-products that are delivered by the Manufacturer to the market segments. The Manufacturer in figure 7-2 aims its platform products at two market regions, each of which has two market segments. Therefore there are four market segments should be served. This also indicates that the maximum number of product variants in the product family is four. With respect to the customer requirements, we assume that:

Assumption 3: The primary requirement of the customers in each market segment can be represented by the required module options of each variant module, and it is only allowed to configure a product that offers higher-ordered module options than the customer's requirements.

Assumption 4: The forecasted demand of a market segment is not affected by the actual PPD decision of the Manufacturer.

With the above Assumption 3 and 4 which imply that the total revenue of the Manufacturer is not affected by the actual PPD decision, minimizing the total cost maximizes the total profit of the Manufacturer. In this chapter, the Manufacturer would choose module options and suppliers simultaneously to minimize the total cost of designing, procuring, ordering, and inventory of module options. As we have mentioned in section 1, platforming, i.e. replacing a lower-ordered module option with a higher-ordered module option, can create economies of scale when developing and ordering module options. However, this must be traded off against the increases in purchasing and inventory cost of module options.

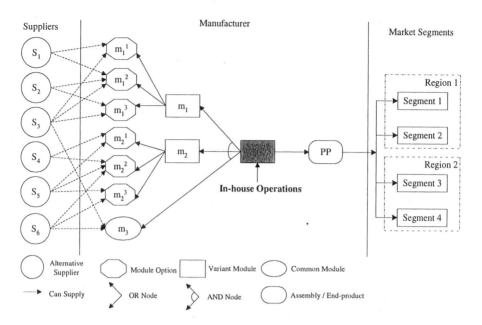

Figure 7-2. Integrated configuration of platform products and supply chain (ICPPSC)

4. NOTATIONS AND MODEL FORMULATION

4.1 Notations

We consider a Manufacturer facing the customer demands of market segment indicated by the subscript $i = 1, 2, ..., I$. The platform products produced by the Manufacturer can be split into a set of modules including variant modules indicated by the subscript $j = 1, 2, ..., J$. Each variant module j can have many module options that could be interchanged to provide the desired performance levels, and are indicated by the superscript $k = 1, 2, ..., K_j$, just like what in figure 7-2. According to Assumption 1, module option k has a higher performance level than module option k' if $k > k'$. All the module options are purchased from a fixed number of alternative suppliers indicated by the subscript $s = 1, 2, ..., S$. Each supplier s can only produce one kind of variant module, namely j^s.

According to Assumption 3, we define the primary requirement matrix, **R**, such that:

$$\mathbf{R}_{j,i} = k,$$

where k is the primary module option of variant module j required by the customers in market segment i. Also, we define the PPD decision matrix of the Manufacturer, **M**, to represent the mix of module options eventually used in the platform products by the Manufacturer, such that:

$$\mathbf{M}_{j,i} = k,$$

where k is the selected module option used for variant module j in the end-product serving market segment i.

Let the Manufacturer's estimate of the lifecycle demand at each market segment be d_i ($i = 1, ..., I$), then the total market demand for the product family, D, is $\sum_{i=1}^{I} d_i$. Let d_{jk} represent the lifecycle demand for variant module j's option k according to customers' requirements, then we have,

$$d_{jk} = \sum_{i=1}^{I} d_i \alpha_{jk}^i, \text{ where } \alpha_{jk}^i = \begin{cases} 1 & \text{if } \mathbf{R}_{j,i} = k \\ 0 & \text{otherwise} \end{cases}.$$

Similarly, let u_{jk} denote the lifecycle demand rate for variant module j's option k according to the actual platform products configuration. Then we have

$$u_{jk} = \sum_{i=1}^{I} d_i \beta_{jk}^i, \text{ where } \beta_{jk}^i = \begin{cases} 1 & \text{if } \mathbf{M}_{j,i} = k \\ 0 & \text{otherwise} \end{cases}.$$

Table 7-1. Parameters and variables of the Manufacturer

Symbol	Meaning
v_{jk}	Binary decision variable to indicate whether module option k is used by the Manufacturer
DC_{jk}	Fixed cost of designing, prototyping, and testing module option k
h	Holding cost rate per unit per year at the Manufacturer
A	Ordering cost occurs at each order placed by the Manufacturer
p_{jks}	Unit selling price for module option k purchased from supplier s

From a set of alternative suppliers S_{jk} who are capable of producing the module option k of variant module j, the Manufacturer awards u_{jks} to supplier s. We define SCC decision matrix $\mathbf{U}$ to represent the sourcing decision result of the Manufacturer, such that:

$$\mathbf{U}_{j,k,s} = u_{jks},$$

where $s \in S_{jk}$.

We also define a SCC decision vector $\mathbf{T}$ to represent the ordering decision of the Manufacturer, such that:

$$\mathbf{T} = (T_1, \ldots T_s, \ldots, T_S),$$

where T_s is the Manufacturer's decision of order interval to supplier s. If supplier s is used by the Manufacturer to supply some module options, then T_s exists and is larger than zero; otherwise, T_s is null.

Other relevant notations are designed in Table 7-1. Note that the v_{jk} in this table can be deduced from the PPD decision matrix $\mathbf{M}$.

4.2 Model formulation

When making PPD and SCC decisions, the Manufacturer incurs two different costs, namely development cost (*DC*), and souring cost (*SC*). The sourcing cost is associated with purchasing, ordering and inventory of module options that make up the platform products. Once the configuration decision of the platform products, or PPD decision matrix $\mathbf{M}$, is given, the development cost *DC* becomes a fixed cost based on the selected module options. In other words, *DC* is decided by the Manufacturer's PPD decision. The sourcing cost *SC*, on the other hand, is a variable cost affected by the Manufacturer's SCC decision.

Then the problem of ICPPSC that minimizes the total cost paid by the Manufacturer under the planning horizon can be formulated as the following constrained optimization problem:

ICPPSCP:

$$\underset{\mathbf{M},\mathbf{U},\mathbf{T}}{Min} \sum_{j=1}^{J} \sum_{k=1}^{K_j} DC_{jk} v_{jk} + \sum_{s=1}^{S} \left[\sum_{j=j^s}^{j^s} \sum_{k=1}^{K_j} u_{jks} p_{jks} + \frac{A}{T_s} + \frac{h U_s T_s}{2} \right]$$

subject to:

$$\mathbf{M}_{j,i} \geq \mathbf{R}_{j,i}, \text{ for } i = 1, 2, \ldots, I; j = 1, 2, \ldots, J, \tag{1}$$

$$v_{jk} = \begin{cases} 1 & \text{if } u_{jk} > 0 \\ 0 & \text{otherwise} \end{cases}, \text{ for } i = 1, 2, ..., I; j = 1, 2, ..., J \ , \tag{2}$$

$$U_s = \sum_{j=j^s}^{j^s} \sum_{k=1}^{K_j} u_{jks} \ . \tag{3}$$

Constraint (1) ensures Assumption 1 of one-way substitution. Constraint (2) sets the value of v_{jk}. Constraint (3) sets the value of U_s which represents the total unit of module options purchased from the capable supplier s. As we have assumed that each supplier is only capable of producing one kind of variant module, all the module options purchased from the same capable supplier s are therefore belonging to one variant module, j^s.

The objective function includes two terms. The first term is the development cost DC. The second term is the sourcing cost SC, which is in turn composed of three parts. The first part is total unit purchasing cost ($PurC$) paid to suppliers, the second part is ordering cost ($OrdC$), and the last one is raw material inventory holding cost ($InvC$). Since DC becomes constant once the Manufacturer's PPD decision is given, the Manufacturer's objective function after he made its PPD decision is SC:

$$\begin{aligned} SC(\mathbf{U,T}) &= PurC + OrdC + InvC \\ &= \sum_{s=1}^{S} \left[\sum_{j=j^s}^{j^s} \sum_{k=1}^{K_j} u_{jks} p_{jks} + \frac{A}{T_s} + \frac{hU_s T_s}{2} \right] \end{aligned} \tag{4}$$

The relevant sourcing cost of the Manufacturer incurred by the procurement of module options from each supplier $s = 1, 2, ..., S$, namely SC_s, can be represented by:

$$\begin{aligned} SC_s(\mathbf{U}_{..s}, T_s) &= PurC_s + OrdC_s + InvC_s \\ &= \sum_{j=j^s}^{j^s} \sum_{k=1}^{K_j} u_{jks} p_{jks} + \frac{A}{T_s} + \frac{hU_s T_s}{2}, \end{aligned} \tag{5}$$

and $\sum_{s=1}^{s} SC_s = SC$.

4.3   Revised model

We revise the objective function of the above ICPPSCP model to a more convenient form for analysis and computational purposes. First, let us consider the optimal SCC decision vector, $\mathbf{T}$, of the Manufacturer.

The optimal ordering interval of the Manufacturer to supplier s, T_s, to minimize SC_s can be obtained following the classical Economic Order Quantity (EOQ) policy. Through equation (5), we have

$$T_s^* = \sqrt{\frac{2A}{hU_s}}.$$

Substituting the above T_s^* into equation (4), we can rewrite it as follows:

$$SC(\mathbf{U}) = \sum_{s=1}^{S}\left[\sum_{j=j^s}^{j^s}\sum_{k=1}^{K_j} u_{jks}\, p_{jks} + \sqrt{2AhU_s}\right].$$

Therefore, the objective function of the ICPPSCP model can be revised to the follows:

ICPPSCP':

$$\operatorname*{Min}_{M,U}\ \sum_{j=1}^{J}\sum_{k=1}^{K_j} DC_{jk} v_{jk} + \sum_{s=1}^{S}\left[\sum_{j=j^s}^{j^s}\sum_{k=1}^{K_j} u_{jks}\, p_{jks} + \sqrt{2AhU_s}\right],$$

subject to:

Constraints (1)-(3).

5. SOLUTION PROCEDURE

As it is reasonable to assume that the platform products have a finite number of variant modules, module options, and capable suppliers, the Manufacturer's feasible PPD and SCC decisions are finite. Therefore the optimal solution to the above ICPPSCP' model can be found by enumeration. However, the enumerative algorithm may be very inefficient because of the high computational complexity. In this section, in order to simplify the feasible solution space and improve the efficiency of the solution algorithm, we begin with deriving two theorems that allow us to rewrite the formulation of the ICPPSCP' model in a more convenient form.

First we derive a theorem that can simplify the sourcing decision matrix, **U**, of the Manufacturer.

Theorem 1: In the optimal SCC decisions of the Manufacturer in our ICPPSCP' model, either all or none of the demand of a module option k is awarded to exactly one capable supplier s.

A formal proof of the above theorem is provided in Appendix A, which is based on the idea that the cost of sourcing a module option from exactly one capable supplier, or single sourcing, is lower than sourcing it from more than one capable supplier, or multiple sourcing. Theorem 1 implicates that if a module option is selected by the Manufacturer in its PPD decision, the Manufacturer will use single sourcing strategy when procuring this module option. Therefore, we refer to this property of the ICPPSCP' model as the *single sourcing* property.

The *single sourcing* property (Theorem 1) simplifies the Manufacturer's supplier selection decision because only the condition of single sourcing needs to be considered for each module option. It allows us to rewrite the Manufacturer's three-dimensional sourcing decision matrix $\mathbf{U}$ into a two-dimensional decision matrix, namely supplier selection matrix Y, to represent the selected supplier for each module option, such that:

$$\mathbf{Y}_{j,k} = s, \ s \in S_{jk}.$$

Then the term *SC* in the objective function of the above ICPPSCP' model can be revised as follows:

$$SC(\mathbf{Y}) = \sum_{s=1}^{S} \left[\sum_{j=j^s}^{j^s} \sum_{k=1}^{K_j} u_{jk} y_{jks} p_{jks} + \sqrt{2AhU_s} \right], \tag{6}$$

where

$$y_{jks} = \begin{cases} 1 & \text{if } \mathbf{Y}_{j,k} = s \\ 0 & \text{otherwise} \end{cases}, \text{ and} \tag{7}$$

$$U_s = \sum_{k=1}^{K_j} u_{jk} y_{jks}. \tag{8}$$

Therefore the objective function of the ICPPSCP' model can be revised to the follows:

ICPPSCP'':

$$\underset{\mathbf{M},\mathbf{U}}{Min} \sum_{j=1}^{J}\sum_{k=1}^{K_j} DC_{jk} v_{jk} + \sum_{s=1}^{S}\left[\sum_{j=j^s}^{j^s}\sum_{k=1}^{K_j} u_{jk}\, y_{jks}\, p_{jks} + \sqrt{2AhU_s} \right],\qquad(9)$$

subject to:

Constraints (1)-(2), (7)-(8).

Next, we derive the second theorem, which is based on the *single sourcing* property (Theorem 1), and can simplify the PPD decision matrix, **M**, of the Manufacturer.

Theorem 2: In the optimal PPD decision of the Manufacturer, either all or none of the demand of a module option k is replaced by a higher performance module option b.

A formal proof is presented in Appendix B, which is based on the idea that the cost of either completely procuring or completely replacing a module option is lower than both procuring and replacing it. Theorem 2 has two implications: (1) a module option k is either fully replaced by a module option with higher performance or is fully outsourced, and (2) if module option k is replaced, exactly one higher performance module option will replace it. This property, according to which a module option is never partially replaced by a higher performance module option, is referred to by us as the *absolute replacement* property of the ICPPSCP' model.

The *absolute replacement* property (Theorem 2) simplifies the Manufactuer's PPD decision. Let F' denote the original number of feasible PPD decision matrix **M** in the ICPPSCP' model. Then it can be written as:

$$F' = \prod_{j=1}^{J}\prod_{i=1}^{I}(K_j - \mathbf{R}_{j,i} + 1)\cdot$$

Through the *absolute replacement* property (Theorem 2), the value of F' can be reduced to F which can be written as:

$$F = \prod_{j=1}^{J}\prod_{i=1}^{I}(K_j - \mathbf{R}_{j,i} + 1) \left/ \prod_{j=1}^{J}\prod_{i=1,\prod_{t=1}^{i-1}(R_{j,i}-R_{j,t})=0}^{i-1}(K_j - \mathbf{R}_{j,i} + 1) \right.$$

Let's indicate the feasible decision matrix $\mathbf{M}$ by superscript $f = 1, 2, ..., F$, and indicate the feasible decision matrix $\mathbf{Y}$ by superscript $n = 1, 2, ..., N$, where $N = \prod_{j=1}^{J} \prod_{k=1}^{K_j} S_{jk}$. Then the number of feasible decisions of the ICPPSCP" model is $F \times N$.

In this chapter, we use a two-level iterative procedure to find the optimal solution of the ICPPSCP" model. For each $\mathbf{M}^f$ (the first level iteration), we find the optimal $\mathbf{Y}^n$ (the second level iteration). After $F \times N$ iterations, the optimal PPD and SCC decisions will be found. Figure 7-3 summarizes the solution procedure.

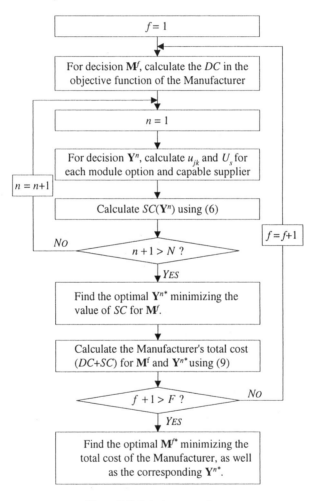

Figure 7-3. Solution procedure

6.   A NUMERICAL EXAMPLE

In this section we present a very simple numerical example to demonstrate the applicability of the proposed solution procedure to our problem.

We aim to construct a group of platform products which retail at approximately \$1,000. These platform products have two variant modules (m_1 and m_2) in their GBOM. Each of the two variant modules has 4 module options. The relative values of m_1 and m_2 to the total value of an entire product are approximately at 18.3% and 11.8% respectively.

The product family serves 2 market regions, namely Europe (EU) and North America (NA), and each regional market has 4 market segments. Thus there are totally 8 market segments. The lifetime demand volume of the platform products is assumed to be 500000 units. It is reasonable that each market region has a different market size. Thus, we assume that EU and NA have approximately 37 and 63 percent of the given worldwide demand volume (D) respectively. Demand volume for each market segment is equal to the average demand volume of the market segments of the region. With respect to primary requirements of market segments, we assume that EU prefers lower-end module options while UA prefers higher-end module options. The primary requirement matrix, **R**, is generated randomly for our simulations as follows:

$$\mathbf{R} = \begin{pmatrix} 2 & 1 & 1 & 2 & 3 & 4 & 3 & 4 \\ 1 & 3 & 1 & 2 & 4 & 2 & 3 & 3 \end{pmatrix}.$$

The development cost of an entire end-product is about \$3 millions and the development cost of each variant module is proportionately determined according to its relative value. Further, it is reasonable that developing a higher-end module option often costs more than developing a lower-end module option. Therefore, we increase development cost for each module option by 5 percent for one performance level higher.

The Cost of Good Sold (COGS) of a finished product is estimated at about 70 percent of its retail price, while about 70 to 75 percent of the COGS is budgeted for raw material cost by the Manufacturer. The retail price has been set at \$1,000 as above, the total raw material cost of a finished product is about \$490. The material cost of each variant module is proportionately determined according to its relative value, and increases by 5 percent for one performance level higher. The ordering cost is set at \$300 per order and the holding cost rate is \$4 per unit per year.

As we have stated before, there are two kinds of suppliers for each variant module, i.e. the supplier with high flexibility which can produce all the module options and the supplier with low flexibility which can only produce either the lower or the higher half module options. We assign three suppliers to each variant module, of which two are with low flexibility and one is with high flexibility. It is reasonable to assume that the prices of the module options from the low-flexibility suppliers are lower than those from the high-flexibility suppliers, so that we set the price of the same module option charged by the low-flexibility suppliers 0.06 percent lower than the price charged by the high-flexibility suppliers.

After the lifetime of the studied platform products described above is assumed as two years, the optimal solutions of the ICPPSC problem are searched through the proposed solution procedure. We present the results in the following.

Table 7-2 shows the comparison between the primary requirements of customers or the case where platforming is not allowed (namely "no platforming" or NP case) therefore the Manufacturer takes the primary customer requirements as its PPD decision, and the optimal PPD decision gotten by the Manufacturer when he decides to use platforming strategy (namely "with platforming" or WP case). In this table, the first and the second columns list modules and module options respectively. In the next eight columns, ■ means the module option selection according to the primary requirements of customers, or in NP case, and ○ means the optimal module option selection in WP case.

As shown in Table 7-2, products in some market segments are diverted, and it results in the decreasing of the number of module options used. More specifically, for each variant module, 4 module options are used in NP case, while only 3 module options are used in WP case. The decreases in the numbers of different module options lead to a decrease of the level of customization of the platform products, as well as a reduction of the development cost from NP case to WP case.

Table 7-2. Platform products configuration results

Module	Option	Region 1				Region 2				# of kinds
		P_1	P_2	P_3	P_4	P_5	P_6	P_7	P_8	
m_1	m_1^1		◘	◘						
	m_1^2	■			■					
	m_1^3	○			○	◘		◘		
	m_1^4						◘		◘	4→3
m_2	m_2^1	◘		◘						
	m_2^2				■		■			
	m_2^3		◘		○		○	◘	◘	
	m_2^4					◘				4→3

■ : Selected in NP case; ○ : Selected in WP case.

Table 7-3 shows the optimal SCC decisions of the Manufacturer in NP and WP case respectively. The selected supplier for each used module option in each case is indicated by the symbol "√" in the table. The total demand of module options awarded and the order interval to the selected supplier are also given in this table. We can see from this table that suppliers with low flexibilities are selected in WP case while in NP case the supplier with high flexibility is used.

Table 7-4 presents the various cost components of the optimal SCC decision in NP case and the optimal ICPPSC decision in WP case. We can see that platforming strategy contributes to the noticeable reduction of $269,189.6 in the total cost of the Manufacturer as well as to the reductions in the development cost *DC*.

Table 7-3. Supply chain configuration results

Case	Sup.	m_1^1	m_1^2	m_1^3	m_1^4	m_2^1	m_2^2	m_2^3	m_2^4	U_s	T_s
NP	S_1	√	√							185000	0.020
	S_2			√	√					315000	0.015
	S_3										
	S_4										
	S_5										
	S_6					√	√	√	√	500000	0.012
WP	S_1	√								92500	0.029
	S_2			√	√					407500	0.014
	S_3										
	S_4					√				92500	0.029
	S_5							√	√	407500	0.014
	S_6										

Table 7-4. Various costs results

		In NP case	In WP case	Reduction
SC	PurC	$74,582,476.8	$75,113,065.1	
	OrdC+InvC	$117,674.8	$130,596.9	
	Total	$74,700,151.6	$75,243,662	-0.73%
	DC	$3,341,100	$2,528,400	24.3%
	Total cost	$78,041,251.6	$77,772,062	0.34%

In conclusion, the computational results indicate that there do have impacts of the Manufacturer's PPD strategies on its SCC decision. Firstly, the total cost of the Manufacturer decreases in WP case. Secondly, the platforming strategy leads the Manufacturer to choose the suppliers with low flexibility in stead of those with high flexibility.

7.   CONCLUSION AND FUTURE WORK

The work presented in this chapter makes several contributions to existing research. First, it adds to existing research theme on the Integrated Configuration of Platform Products and Supply Chain (ICPPSC) problem from the perspective of a single manufacturer and its supply network. In addition, we propose an overall research framework on the ICPPSC problem, according to which, the work presented in this chapter has been conducted. The second important contribution is the formulation of a non-linear mathematical decision model of the specific ICPPSC problem in the scenario of non-interactive supply chain, derivation of analytical properties (*single sourcing* and *absolute replacement*), and development of an iterative approach to find the optimal solution of the problem. The model may be used by production managers as a decision support tool to make decisions such as which end-products to design and produce, which suppliers to use for supplying which module option. Finally, the proposed model and solution procedure are applied to a simple numerical example. Computational results are presented and the indications are discussed. The use of platform commonality and modularity strategies has been found beneficial to the manufacturer's supply network and allows the manufacturer to choose suppliers with low flexibility or capability.

There are several aspects which merit further research attention. An immediate extension is to conduct a more complex case study to obtain more understanding and knowledge about the mutual impacts of the PPD and SCC decisions. The results will be compared with those obtained in our previous study (Huang et al., 2005b) where the suppliers are followers in a dynamic non-cooperative game with the manufacturing firm as the leader.

Based on the optimality properties derived in this work, the current solution procedure is enumerative and therefore has a comparatively low computational efficiency. A more sophisticated solution algorithm is required to solve the problem for better efficiency and effectiveness. We plan to develop a solution algorithm based on the Genetic Algorithm for solving the model.

In the meantime, we would like to relax some assumptions of the research problem studied in this chapter, such as Assumption 2 that the capacity of each supplier is unlimited, to improve its generality. On the other hand, we would incorporate more PPD and SCC parameters and decision variables that are sensitive to platform commonality and modularity strategies to increase the complexity of the mathematical model. In this case, we can get more observations and implications from the experimental simulations.

Except for the three scenarios studied in Huang et al. (2005a, 2005b) and in this chapter, other scenarios should be studied and results should be compared against with each other. Example scenarios include (a) the suppliers are wholly owned subsidiary of the manufacturing firm, (b) the relationship between the manufacturer and the suppliers are cooperative in the supply chain, and (c) the relationship among the suppliers are competitive in the supply chain.

Lastly, it is desirable to extend the study about the two-echelon supply chain with the manufacturer and its suppliers to supply chains with more echelons such as retailers and distributors in future.

ACKNOWLEDGEMENT

The authors are most grateful to the University of Hong Kong Research Committee for the partial financial supports for this research.

APPENDIX A: PROOF OF THEOREM 1

We use contradiction to proof this result. First, let's consider the case when there are two capable suppliers.

Suppose a module option k of variant module j is supplied by suppliers s and s', i.e. $u_{jk} = u_{jks} + u_{jks'}$ and $0 < u_{jks} < u_{jk}$. Without any loss of generality, we assume that $p_{jks'} > p_{jks}$. Let U_s' and $U_{s'}'$ denote the total unit of module options that have been purchased from supplier s and s' by the Manufacturer respectively, except the module option k. The relevant terms of the Manufacturer's objective function can be written as:

$$u_{jks} p_{jks} + \sqrt{2Ah\left(U_s' + u_{jks}\right)} + (u_{jk} - u_{jks}) p_{jks'} + \sqrt{2Ah\left(U_{s'}' + u_{jk} - u_{jks}\right)}$$

$$= u_{jk} p_{jks'} - u_{jks}\left(p_{jks'} - p_{jks}\right) + \sqrt{2Ah}\left(\sqrt{U_s' + u_{jks}} + \sqrt{U_{s'}' + u_{jk} - u_{jks}}\right)$$

Let we define

$$f(u_{jks}) = -u_{jks}\left(p_{jks'} - p_{jks}\right), \text{ and}$$

$$g(u_{jks}) = \sqrt{U_s' + u_{jks}} + \sqrt{U_{s'}' + u_{jk} - u_{jks}}.$$

Then the above relevant terms of the Manufacturer's objective function can be written as

$$u_{jk} p_{jks'} + f(u_{jks}) + \sqrt{2Ah}\, g(u_{jks}) \cdot$$

From $g(u_{jks})$, we can get that $g''(u_{jks}) < 0$ or $g(u_{jks})$ is concave down for all u_{jks} in $[0, U_{s'}' + u_{jk}]$. In addition, it is clear that $g(U_{s'}' + u_{jk}) \le g(u_{jk})$ and $g\left(U_{s'}' + u_{jk}\right) \le g(0)$. Thus the value of $g(u_{jks})$ for all u_{jks} in $[0, U_{s'}' + u_{jk}]$ has a local minimum at $u_{jks} = U_{s'}' + u_{jk}$, which implies the value of $g(u_{jks})$ is the lowest when $u_{jks} \geq u_{jk}$. On the other hand, we note that the value of $f(u_{jks})$ can be lowered by increasing the value of u_{jks}, i.e. the value of $f(u_{jks})$ also being the lowest when $u_{jks} \geq u_{jk}$. Therefore, the value of the above relevant terms of the Manufacturer's objective function is the lowest when $u_{jks} \geq u_{jk}$. This conclusion contradicts the initial assumption that $0 < u_{jks} < u_{jk}$.

Then, in the case where there are more than two capable suppliers who can produce module option k, the similar proof as above can be used between every two capable suppliers. Hence, it is optimal to source the module option k from exactly one capable supplier.

APPENDIX B: PROOF OF THEOREM 2

The proof is in two parts. First, we prove that a module option k is never partially replaced by a higher module option b. Second, we prove that if a module option k is fully replaced, it will be replaced by exactly one higher module option b. We also use contradiction in this proof.

Part I: On the basis of the *single sourcing* property (Theorem 1), we suppose that $0 < u_{jk} < d_{jk}$, and the optimal PPD decision is such that u_{jk} units of module option k are procured from a supplier s and the remaining $d_{jk} - u_{jk}$ units are replaced by the module option b which is procured from a supplier s'. Let U'_s and $U'_{s'}$ denote the total unit of module options that have been purchased from supplier s and s' by the Manufacturer respectively, except the module option k and $d_{jk} - u_{jk}$ units of module option b. The relevant terms of the Manufacturer's objective function can be written as:

$$u_{jk} p_{jks} + \sqrt{2Ah\left(U'_s + u_{jk}\right)} + (d_{jk} - u_{jk}) p_{jbs'} + \sqrt{2Ah\left(U'_{s'} + d_{jk} - u_{jk}\right)}$$
$$= d_{jk} p_{jbs'} - u_{jk}\left(p_{jbs'} - p_{jks}\right) + \sqrt{2Ah}\left(\sqrt{U'_s + u_{jk}} + \sqrt{U'_{s'} + d_{jk} - u_{jk}}\right).$$

According to Assumption 1, we have $p_{jbs'} - p_{jks} > 0$. Let we define

$$f(u_{jk}) = -u_{jk}\left(p_{jbs'} - p_{jks}\right), \text{ and}$$

$$g(u_{jk}) = \sqrt{U'_s + u_{jk}} + \sqrt{U'_{s'} + d_{jk} - u_{jk}}.$$

Then the above relevant terms of the Manufacturer's objective function can be written as

$$d_{jk} p_{jbs'} + f(u_{jk}) + \sqrt{2Ah} g(u_{jk}).$$

Similar with the proof of the *single souring* property (Theorem 1), we can get the conclusion that the value of the above relevant terms of the Manufacturer's objective function is the lowest when $u_{jk} \geq d_{jk}$, which contradicts the initial assumption that $0 < u_{jk} < d_{jk}$. Hence, a module option k is never partially replaced by a higher module option b.

Part II: Suppose that a module option k is replaced by two module options b and c which are procured from supplier s and s' respectively. Without any loss of generality, it can be assumed that $c > b$. Suppose $0 < d'_{jk} < d_{jk}$, and the optimal PPD decision of the Manufacturer is such that module options b and c substitute d'_{jk} and $d_{jk} - d'_{jk}$ units of demand of module option k. Let U'_s and $U'_{s'}$ denote the total unit of module options that have been purchased from supplier s and s' by the Manufacturer respectively, except d'_{jk} units of module option b and $d_{jk} - d'_{jk}$ units of module option c. The relevant terms of the Manufacturer's objective function can be written as:

$$d'_{jk} p_{jbs} + \sqrt{2Ah\left(U'_s + d'_{jk}\right)} + (d_{jk} - d'_{jk}) p_{jcs'} + \sqrt{2Ah\left(U'_{s'} + d_{jk} - d'_{jk}\right)}$$
$$= d_{jk} p_{jcs'} - d'_{jk}\left(p_{jcs'} - p_{jbs}\right) + \sqrt{2Ah}\left(\sqrt{U'_s + d'_{jk}} + \sqrt{U'_{s'} + d_{jk} - d'_{jk}}\right).$$

According to Assumption 1, we have $p_{jcs'} - p_{jbs} > 0$. Let we define

$$f(d'_{jk}) = -d'_{jk}\left(p_{jcs'} - p_{jbs}\right), \text{ and}$$

$$g(d'_{jk}) = \sqrt{U'_s + d'_{jk}} + \sqrt{U'_{s'} + d_{jk} - d'_{jk}}.$$

Then the above relevant terms of the Manufacturer's objective function can be written as

$$d_{jk} p_{jcs'} + f(d'_{jk}) + \sqrt{2Ah}\, g(d'_{jk}).$$

Similarly, we can get the conclusion that the value of the above relevant terms of the Manufacturer's objective function is the lowest when $d'_{jk} \geq d_{jk}$, which contradicts the initial assumption that $0 < d'_{jk} < d_{jk}$. Hence, a module option k is never replaced by more than one higher module option.

REFERENCES

Baker, K. R., 1985, Safety stocks and commonality, *Journal of Operations Management* **6**(1): 13 22.

Benton, W. C., and Krajewski, L. J., 1990, Vendor performance and alternative manufacturing environments, *Decision Sciences* **21**(2): 403 415.

Berry, W. L., Tallon, W. J., and Boe, W. J., 1992, Product structure analysis for the master scheduling of assemble-to-order products, *International Journal of Operations & Production Management* **12**(11): 24 41.

Chakravarty, A. K., and Balakrishnan, N., 2001, Achieving product variety through optimal choice of module variations, *IIE Transactions* **33**: 587 598.

Collier, D. A., 1981, The measurement and operating benefits of component part commonality, *Decision Sciences* **12**(1): 85 96.

Collier, D. A., 1982, Aggregate safety stock levels and component part commonality, *Management Science* **28**(11): 1296 1303.

Dogramaci, A., 1979, Design of common components considering implications of inventory costs and forecasting, *AIIE Transactions* **11**(2): 129 135.

Gerchak, Y., and Henig, M., 1989, Component commonality in assemble-to-order systems: models and properties, *Naval Research Logistics* **36**: 61 68.

Gerchak, Y., Magazine, M. J., and Gamble, A. B., 1988, Component commonality with service level requirements, *Management Science* **34**(6): 753 760.

Gupta, S., and Krishnan, V., 1999, Integrated component and supplier selection for a product family, *Production and Operations Management* **8**(2): 163 182.

Huang, G. Q., Zhang, X. Y., and Liang, L., 2005a, Towards integrated optimal configuration of platform products, manufacturing processes, and supply chains, *Journal of Operations Management* **23**: 267 290.

Huang, G. Q., Zhang, X. Y., and Lo, V. H. Y., 2005b, Integrated configuration of platform products and supply chains for mass customization: A game-theoretic approach, *IEEE Transactions on Engineering Management*, submitted for publication.

Meyer, M. H., 1997, Revitalize your product lines through continuous platform renewal, *Research-Technology Management* **40**(2): 17 28.

Meyer, M. H., and Lehnerd, A. P., 1997, *The Power of Product Platforms: Building Values and Cost Leadership*, The Free Press, New York.

Park, B. J., 2001, *A framework for integrating product platform development with global supply chain configuration*, GIT PhD Dissertation, Georgia.

Robertson, D., and Ulrich, K., 1998, Planning for Product Platforms, *MIT Sloan Management Review* **39**(4): 19 31.

Salvador, F., Forza, C., and Rungtusanatham, M., 2000, How to mass customize: Product architecture, sourcing configurations, *Business Horizons* **45**(4): 62 69.

Salvador, F., Forza, C., and Rungtusanatham, M., 2002, Modularity, product variety, production volume, and component sourcing: theorizing beyond generic prescriptions, *Journal of Operations Management*, **20**: 549 575.

Ulrich, K. T., and Eppinger, S. D., 1995, *Product Design and Development*, McGraw-Hill, New York.

Wheelwright, S. C., and Clark, K. B., 1992, *Revolutionizing Product Development – Quantum Leaps in Speed, Efficiency and Quality*, The Free Press, New York.

Chapter 8

MODULARITY AND DELAYED PRODUCT DIFFERENTIATION IN ASSEMBLE-TO-ORDER SYSTEMS
Analysis and Extensions from a Complexity Perspective

Thorsten Blecker and Nizar Abdelkafi
Hamburg University of Technology, Department of Business Logistics and General Management

Abstract: The paper assumes a product design around modular architectures and discusses the suitability of the principle of delayed product differentiation in assemble-to-order environments. We demonstrate that this principle does not enable one to make optimal decisions concerning how variety should proliferate in the assembly process. Therefore, we propose to complement this principle in that we additionally consider the variety induced complexity throughout the assembly process. The weighted Shannon entropy is proposed as a measure for the evaluation of this complexity. Our results show that the delayed product differentiation principle is reliable when the selection probabilities of module variants at each assembly stage are equal and the pace at which value is added in the whole assembly process is constant. Otherwise, the proposed measure provides different results. Furthermore, the entropy measure provides interesting clues concerning eventual reversals of assembly sequences and supports decisions regarding what modules in an assembly stage could be substituted by a common module.

Key words: Modularity, delayed product differentiation, complexity, weighted Shannon entropy

1. INTRODUCTION

Assemble-to-order is a business model whereby final product variants are not assembled until customer order arrives. It can be considered as one form of practicing mass customization because the products are individualized out of components, which are held in a generic form. If these components can be combined in very different ways, a large product variety would be triggered, thereby increasing the complexity of operations. The negative effects of product variety and complexity on both efficiency and responsiveness are well-known and have already been discussed by many authors (e.g. Blecker et al. 2005). To alleviate the negative impacts of variety and complexity, postponement and delayed product differentiation are proposed as suitable strategies.

In this chapter, we assume a modular product architecture, which means that product variations are obtained by mixing and matching a set of modules with well defined interfaces. This assumption can be seen from two different perspectives with respect to the principle of delayed product differentiation. In effect, we can interpret both concepts to be related to each other and consider delayed differentiation as a natural consequence of the use of modules. This interpretation is justified in that modules are held at a generic form and that their assignment to different variations is deferred until concrete demand is available. However, we can view delayed product differentiation from anther perspective, which aims at minimizing variety proliferation throughout the process of final assembly. This perspective focuses on keeping the number of different subassemblies in the process at a low level. In our discussion, we especially deal with the second interpretation. We will show that this interpretation can lead to suboptimal results in assembly-to-order environments. Therefore, it should be complemented by a second principle which is called the principle of minimum variety-induced complexity.

The next section provides a short literature review on modularity, postponement, delayed product differentiation, and complexity. In section 3, we deal with the insufficiencies of the delayed product differentiation principle in assemble-to-order systems. In section 4 we introduce the principle of minimum variety induced complexity and present its theoretical background. We also explore its application in a two-stage assembly process. Finally, section 5 concludes and presents directions for future research.

2.   LITERATURE REVIEW

2.1   Modularity

In the technical literature, there are numerous definitions of the term "modularity", of which we quote some selected ones. Schilling (2003, p. 172) defines modularity "… as a general systems concept: it is a continuum describing the degree to which a system's components can be separated and recombined, and it refers both to the tightness of coupling between components and the degree to which the "rules" of the system architecture enable (or prohibit) the mixing and matching of components". Whereas Schilling considers modularity in the general case without restrictions concerning the kind of system, Baldwin/Clark particularly focus on products and processes. They define modularity as "building a complex product or process from smaller subsystems that can be designed independently yet function together as a whole" (Baldwin/Clark 2003, p.149). In the context of product archi-tectures, Ulrich (2003, p. 121) points out that "[a] modular architecture in-cludes a one-to-one mapping from functional elements in the function structure to the physical components of the product, and specifies de-coupled interfaces between components". For the purpose of our work, we define modularity as an attribute of the product system that characterizes the ability to mix and match independent and interchangeable product building blocks with standardized interfaces in order to create product variants. The bijective mapping between functional elements and physical building blocks is prefer-able and refers to an extreme and ideal form of modularity.

An important advantage of product modularity is that it enables the pro-duction of large product variety while maintaining low costs. This makes modularity attractive for large variety environments such as mass customi-zation. Efficiency can be achieved due to the economies of scale, economies of scope, and economies of substitution. The economies of scale result from the components rather than products, while the economies of scope arise through the multiple use of a few components in a large number of product variations (Pine 1993). In addition, a modular design permits a partial reten-tion of components when it is to upgrade or improve the performance of the modular system. The costs that are saved because the system is not designed afresh are referred to as economies of substitution (Garud/Kumaraswamy 2003). From an operations' perspective, Duray et al. (2000) point out that modularity is a basic component in manufacturing situations considered to be flexible. It also shortens delivery times because final product configura-tion occurs out of modules made to stock and with high work content. Fur-thermore, since modules are self-contained and have standardized interfaces, they can be manufactured simultaneously and independently of each other, thereby reducing the total production time (Ericsson/Erixon 1999). Modu-larity is generally discussed in connection with delayed product differentia-tion and postponement. Both concepts found increasing popularity in aca-demia and practice, especially when it is to discuss mass customization and assemble-to-order.

2.2   Delayed Product Differentiation and Postponement

Postponement is originally introduced by Alderson (1950) as a concept that reduces risk and uncertainty costs. Bucklin (1965) makes the distinction between three types of postponement, which are time, place, and form postponement. Time postponement refers to the delay of forward shipment of goods, whereas place postponement aims at maintaining goods at central locations in the channel. Form postponement is related to the differentiation of the product itself. Zinn/Bowersox (1988) define four types of form postponement, which are labeling, packaging, assembly and manufacturing postponement.

Within the context of the supply chain, van Hoek (2001, p. 161) defines postponement "… as an organizational concept whereby some of the activities in the supply chain are not performed until customer orders are received. Companies can then finalize the output in accordance with customer preferences and even customize their products". Christopher (2005, p. 134) refers to postponement "…as the process by which the commitment of a product to its final form or location is delayed for as long as possible." In our work, since we will focus on the product and its assembly process, we are not concerned with time and place postponements which are of value when it is to consider the whole supply chain. Therefore, our interest will be only given to form postponement which is in accordance with the delayed product differentiation principle. The main objective of this principle is to delay downstream the activities that are responsible for providing the product an identity according to customer specifications. Theoretically, delayed differentiation involves two parts in the value chain. The first part is production-driven (push system), whereas the second part is customer-driven (pull system). The point in the value chain that separates between both systems is generally called the decoupling point[1]. Lampel/Mintzberg (1996) provides a continuum of strategies concerning the degree of customization and possible locations of this point. Their framework combines customization and standardization whereby the degree of customization decreases as the decoupling point moves downstream in the value chain.

Many authors argue that deferring the stage at which products assume their unique identities considerably reduces the negative impacts of variety on manufacturing performance (e.g. Lee/Tang 1997). Consequently, redesign activities with the objective of delaying product differentiation lead to the achievement of large product variety at low costs. This also is necessary in order to make mass customization work efficiently. In addition, delayed

---

[1]   The decoupling point is sometimes referred to as CODP which stands for Customer Order Decoupling Point (Van Hoek 1997) or OOP which the abbreviation of Order Penetration Point (Sharman 1984).

product differentiation is regarded as an important principle for the reduction of complexity in operations. Since we intend to discuss the suitability of this principle in assemble-to-order environments from a complexity perspective, it is necessary to define at first what we understand under the term "complexity". Therefore, the main purpose of the next section is not to explain the potential of modularity or delayed differentiation in reducing complexity but to provide a suitable definition of complexity to be used throughout this paper.

2.3   Complexity

Complexity is a widely discussed topic in many research fields in science. There are also many attempts to provide a universal and generally admitted definition of complexity. However, a single and generally accepted definition does not exist. Therefore, it is suitable to define complexity in the context of our research field. Since this work can be assigned both to business administration and engineering management, we retain two definitions that are frequently used to deal with research topics in these fields. The first definition describes complexity as an attribute of a system (system theoretical approach). The second one considers complexity as the entropy of a system. In the following, we shall briefly discuss both approaches:

- Complexity from a system theoretical approach:

A system consists of elements or parts (objects, systems of lower order, subsystems) which are connected to each other through relations. To assess complexity, the system elements and relations should be evaluated according to three variables which are: the number, diversity, and states' variety. In effect, the higher the number of the system elements and their relations, the less straightforward is the system, thus resulting in higher complexity. It is noteworthy that the addition of an element to the system leads to a disproportionate increase of the potential relations between the system elements. On the other hand, diversity refers to the homogeneity or heterogeneity of the elements and their relations. It is obvious that the less homogeneous (more heterogeneous) the system elements are, the higher is the system complexity. The third variable which is the states' variety evaluates the instability of the system and indicates its dynamical behavior in the course of time. In other words, as the number and types of the system elements and relations tend to change rapidly, the complexity of the system gets higher (e.g. Ashby 1957, Bertalanffy 1976). We can notice that the system theoretical approach does not provide only one measure that assesses complexity. It is rather based on many dimensions, which constitutes, in fact, its major drawback.

- Complexity as the entropy of a system

Complexity, uncertainty and information are linked to each other. In effect, in order to reduce the complexity of a system, we can simplify it by allowing some degree of uncertainty in its description. This information loss that is necessary for reducing the complexity of the system to a manageable level is expressed in uncertainty (Klir/Folger 1988). As uncertainty grows, the system is more complex since more information is required to describe and monitor each of its states (Sivadasan et al. 2005). In this context, a suitable measure of the uncertainty of a system is the entropy that is introduced by Shannon (1948). In his seminal work, Shannon posed the question: "Can we find a measure of how much "choice" is involved in the selection of the event or how uncertain we are of the outcome?" Then, Shannon (1948) has set forth the following properties to be satisfied by the function $H(p_1,...,p_n)$ where $p_1,...,p_n$ are the probabilities of occurrence of events $1,...,n$:

1. H should be continuous in p_i.
2. If all p_i are equal, $p_i=1/n_i$, then H should be a monotonic increasing function of n. With equally likely events there is more choice, or uncertainty, when there are more possible events.
3. If a choice can be broken down into successive choices, the original H should be the weighted sum of the individual values of H.

Shannon (1948) has demonstrated that the only function that is satisfying the three above assumptions is of the form: $H = -K\sum_{i=1}^{n} p_i \log p_i$ whereby the constant K merely amounts to a choice of a unit of measure. Then Shannon defined entropy of the set of probabilities $p_1,...,p_n$ as $H = -\sum_{i=1}^{n} p_i \log p_i$.

Thus, the value of uncertainty and subsequently complexity of a system taking n states with probabilities $p_1,...,p_n$ can be measured by the entropy function. Due to property 2, the higher the number of states the system can take and the more likely these states tend to occur with the same probability, the higher is the complexity of the system. Intuitively, Feynman (1991) describes the notion of entropy as a measure of disorder that is the number of ways by which the insides of a system (e.g. gas molecules) can be arranged, while from outside it looks the same. As the number of microstates (insides) assigned to a specific macro-state (outside) increases, disorder and subsequently complexity increases. The main advantage of entropy is that it provides a quantitative measure for complexity. We will use entropy later on to evaluate the variety induced complexity in assemble-to-order systems.

3. PROBLEM DESCRIPTION

3.1 Delayed Product Differentiation Principle – An Example

Consider a portion of the assembly process of a Personal Computer (PC). PCs have a modular architecture, in which the following components; processor, motherboard, working memory, graphic card, sound card and hard drive can be considered as independent modules. In effect, each component performs a specific function and has specified interfaces to the motherboard which is the basic component. Furthermore, each module can have many variants. For example, processor variants can be differentiated according to their corresponding frequencies, so that two processors with respective frequencies of 2.0 GHz and 2.3 GHz are two different variants. Due to the modular product architecture of a PC, it is possible to state that the variants of each module are assembled at one sub-process. In addition, suppose that there are no sequencing constraints in the assembly process of the modules mentioned above. The delayed product differentiation principle suggests that variety proliferation should be kept at a low level. In other words, the increase of variety from one sub-process to another should be maintained at a minimum level. To illustrate this, we assume that we have one motherboard type, one hard drive type, 2 processors, 4 graphic card variants, 3 working memory types and 3 sound cards. According to the delayed product differentiation principle, the optimal assembly sequence would be to start from the basic module: motherboard and then to assemble successively the hard drive, processor, working memory or sound card, and finally the graphic card (Figure 8-1).

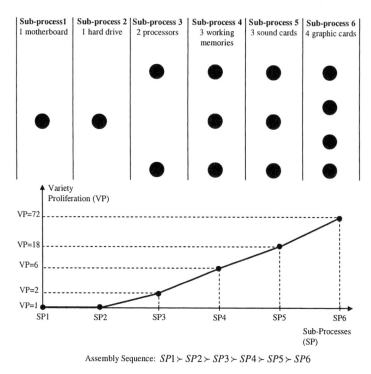

Figure 8-1. Optimal variety proliferation of a PC assembly process according to the delayed product differentiation principle

Note that the curve outlining the increase of variety in figure 8-1 is plotted according to a logarithmic scale in order to enable the representation of a high number of variations. By sequencing the assembly process as it is shown above, it is possible to achieve the lowest variety proliferation. In effect, sub-process 2 triggers no increase of the number of variants in the process since only one hard drive can be assembled to the motherboard. At sub-process 3, two types of processors can be built on the sub-assembly that is made out of the motherboard and hard drive. Mixing and matching modules to each other at the different sub-processes would trigger six possible sub-assembly variations at sub-process 4, 18 possible variations at sub-process 5 and 72 possible variants at the last sub-process. Thus, the flexibility that is ensured by modular product architectures can bring about an exponential increase of variety during the assembly process.

3.2   Insufficiencies of the Delayed Product Differentiation Principle

Now suppose that because of sequencing or assembly process constraints only two possible sequences 1 and 2 can be realized as it is shown by figure 8-2. While assembly process 1 triggers lower variety proliferation than assembly process 2 at the beginning, it exhibits higher proliferation of variety at the end of the process. Thus, we are in front of a situation, in which it is difficult to make a choice between the two possible sequences. On the basis of the delayed product differentiation principle it is not possible to compare between both processes. It cannot provide us with interesting information for optimal decision making. The question mainly concerns if it is better to let variety increase at the beginning of the process, while profiting from decreasing variety at the end of the process or to guarantee low variety at the beginning, while accepting higher proliferation of variety at the end of the process.

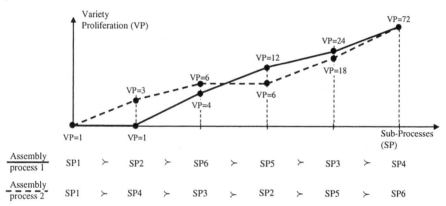

Figure 8-2. Variety proliferation according to two different PC assembly sequences

By means of this simple example, we demonstrate that due to the modularity of products, the delayed product differentiation principle is not sufficient to make optimal decisions concerning the sequence that optimizes variety proliferation. It is worth noting that the problem described may not be available when products are not developed on the basis of modules. In the absence of modularity, mixing and matching components to configure different product variants can be restrained because of incompatible interfaces. As a result, variety proliferation can be strongly constrained and alternative assembly sequences may not be available. In this case, only product redesigns would generate delayed variety proliferation. However, since we intend to examine the relationships between delayed differentiation and modularity in the case of assemble-to-order and mass customization, it is legitimate to assume that the product modularity condition is satisfied.

4.   A NECESSARY COMPLEMENT TO THE DELAYED PRODUCT DIFFERENTIATION PRINCIPLE IN AN ASSEMBLE-TO-ORDER ENVIRONMENT

In this section, we propose to provide a complement to the principle of delayed product differentiation. The main objective is to find the principle(s) that should be additionally taken into account in order to well found decisions concerning variety proliferation. Note that we do not disapprove the principle of delayed differentiation but we have to look for other principles that should be considered in situations when this principle does not support optimal decision making. To achieve this goal, we have to examine at first the reasons that make delayed differentiation insufficient to handle the problem of variety proliferation in assemble-to-order environments.

4.1   Reasons Explaining the Insufficiencies of the Delayed Product Differentiation Principle in an Assemble- To-Order Environment

We shall explore the consequences that result from variety proliferation in make-to-stock (Push) and assemble-to-order (pull) systems. A push production system triggers inventories of components or modules, semi-finished and finished products. Thus, variety proliferation in a push-system brings about an exponential increase of inventory because of high safety stock levels. Since delayed product differentiation reduces variety proliferation, it lowers inventories throughout the production process. For example, Lee (1996) quantitatively demonstrates the potential of delayed product differentiation in decreasing inventories at the finished product level in a make-to-stock environment. The positive consequences of delayed product differentiation can be also seen in the decrease of production planning and scheduling complexity, fewer quality problems, lower purchasing costs, etc.

However, inventories in an assemble-to-order system may be held at the module level but not at the finished product level. It is possible to generate reliable forecasts of the aggregate demand of modules, while postponing final assembly until customer order arrives. Consequently, responsiveness is improved in that delivery times only depend on the assembly lead time, number and work content of waiting orders and shipment time to the customer. It follows that many configurations of the assembly process may involve the same level of inventory at the module level. Therefore, the immediate advantage that results from delayed product differentiation in a push system may not be available in a pull system, thereby making the comparison between two or more assembly sequences difficult as it is shown in section 3.

Recapitulating, the principle of delayed product differentiation can be sufficient in a make-to-stock environment because it reduces the negative impacts of inventories. However, in assemble-to-order systems based on modular product architectures, an additional principle is required. In this context, it is worth noting that the main objective is not to minimize variety proliferation in itself but to optimize performance. An assemble-to-order system is said to be performing well if it provides customers with the required variety[2], while still achieving costs' efficiency and responsiveness. The system performance is however, negatively affected by the complexity that is induced by variety. Martin/Ishii (1996, 1997) determine three indexes for the measurement of what they call "variety complexity": the commonality index, differentiation index, and setup index. The commonality index evaluates the extent to which final products use common components. The differentiation index measures the degree to which variety with high added value and long assembly times proliferates at the end of the process. The setup index compares setup costs to product costs. All three indexes consider the number of variants involved at each assembly stage. It follows because of the reasons explained above that they are of little suitability in assemble-to-order systems. Furthermore, the development of these indexes is not based on an accurate definition of what complexity should be.

We agree with Martin/Ishii (1996, 1997) that the variety induced complexity should be amplified if components or modules with long assembly times and high added values are assembled at the beginning of the process. However, the real complexity effects of variety should be captured by a third variable which is the probability that a module variant would be selected by customers. This variable also enables one to estimate the impacts of commonality and setups. In effect, as the preferences of customers get increasingly polarized along a subset of module variants, the commonality of final products with respect to this subset increases. In addition, the stability of the process flow depends on the number of module variants and their corresponding selection probabilities. Thus, if we succeed in developing a single measure that is based on a precise definition of complexity and that takes the

---

[2]   In this work we are not concerned with the determination of optimal final variety from the customer perspective. This variety is supposed to be given and fulfilled through different module combinations.

selection probabilities, added values, and assembly lead times into account, it would suitable for the evaluation of the variety induced complexity. In the next section, we propose to develop such a measure on the basis of the concept of entropy.

4.2   Model Description and Complexity Measure

For the description of the model, we need the following notations:

n Number of processes in the whole assembly process
j Index of the processes or modules
k Index of the module variants that can be assembled at a process j
M_j The module family that can be assembled at process j

M_{jk} A module variant that can be assembled at process j where $M_j = \{M_{jk}\}_{k=1,..,n_j}$

T_{jk} Assembly time of the module variant M_{jk}
n_j Total number of module variants that can be assembled at process j
v_{jk} Value added due to the assembly of module variant k at process j
p_{jk} Probability of selection of module variant M_{jk}

Furthermore, define $T_j = \sum_{k=1}^{n_j} p_{jk} T_{jk}$ the average assembly time at process j

and $T = \sum_{j=1}^{n} T_j$ the average assembly time of the final products. Similarly, let

$V_j = \sum_{k=1}^{n_j} p_{jk} v_{jk}$ be the average value added at process j and $V = \sum_{j=1}^{n} V_j$ the

average value added in the final products.

In order to define a complexity measure at each process j, we will make use of the weighted Shannon entropy that is defined as (Klir/Folger 1988):

$$H(p(x), w(x) / x \in X) = -\sum_{x \in X} w(x) p(x) \log_2 p(x)$$

where $p(x)$ are probabilities defined on a finite set X and $w(x)$ are weights that are associated with $p(x)$. Note that it is only assumed that weights $w(x)$ are nonnegative and finite real numbers. In the following, we suppose that the assembly lead times of module variants $M_{jk} / k = 1,..,n_j$ at process j are equal, thereby resulting in $T_{jk} = T_j$ for $k = 1,..,n_j$. In other words, it is assumed that the assembly times do not depend on the module variant, but rather on the process (or the module family). This assumption can be justified by the main property of modular products saying that the module interfaces inside a module family M_j are standardized. Define $w_{jk} = \tau_j \delta_{jk}$ where

- $\delta_{jk} = \dfrac{v_{jk}}{V}$ is a coefficient that compares the value added v_{jk} of module M_{jk} to the average value added V in the final products.

- $\tau_j = \dfrac{\sum\limits_{i=j}^{n} T_i}{T}$ is the portion of time that a module M_{jk} spends in the process in comparison to the total lead time required to assemble a final product.

Thus, the expression of w_{jk} is: $w_{jk} = \dfrac{v_{jk}}{V} \dfrac{\sum\limits_{i=j}^{n} T_i}{T}$

The weighted Shannon entropy measure of process j is defined as follows:

$$H_j = -\sum_{k=1}^{n_j} w_{jk} P_{jk} \log_2 P_{jk} = -\frac{1}{T} \sum_{i=j}^{n} T_i \sum_{k=1}^{n_j} \frac{v_{jk}}{V} P_{jk} \log_2 P_{jk}$$

The total entropy of the whole assembly process is the sum of the entropies generated by each process j:

$$H = \sum_{j=1}^{n} H_j = -\frac{1}{VT} \sum_{j=1}^{n} \sum_{i=j}^{n} \sum_{k=1}^{n_j} T_i v_{jk} P_{jk} \log_2 P_{jk}$$

On the basis of the total entropy measure, it is possible to evaluate alternative assembly sequences. The optimal sequence is the one with the lowest variety induced complexity value. Note that we did not consider assembly constraints, which may make the implementation of the optimal solution impossible. However, the entropy measure does not lose its value, since it enables one to choose the next best solution which is the sequence with the next lowest variety induced complexity value. Such a measure can be also seen as the driver that initiates design changes on the product level in order to reduce variety induced complexity of the assembly process. Therefore, it can be seen as a measure that evaluates Design For Assembly (DFA) efforts.

4.3   Exploration of the Complexity Measure for a Two-Stage Assembly Process

In order to illustrate the application of the complexity measure and to gain insights when the delayed product differentiation principle may provide good results and when it fails, we consider an assembly process consisting of two assembly stages A and B. At stage A, n_1 module variants can be assembled on a basic component. At stage B; there are n_2 module variants that can be built on the sub-assemblies coming through stage A. Thus, each final product consists of the basic module; a module variant from stage A and a module variant from stage B (Figure 8-3). Recall that the delayed differentiation principle suggests placing stage A prior to stage B if $n_1 \leq n_2$.

The total weighted entropies of the sequence A-B and sequence B-A are provided by the following expressions:

$$H_{A-B} = -\frac{1}{V}\sum_{k=1}^{n_1} v_{Ak} p_{Ak} \log_2 p_{Ak} - \frac{T_B}{VT}\sum_{k=1}^{n_2} v_{Bk} p_{Bk} \log_2 p_{Bk}$$

$$H_{B-A} = -\frac{1}{V}\sum_{k=1}^{n_2} v_{Bk} p_{Bk} \log_2 p_{Bk} - \frac{T_A}{VT}\sum_{k=1}^{n_1} v_{Ak} p_{Bk} \log_2 p_{Ak}$$

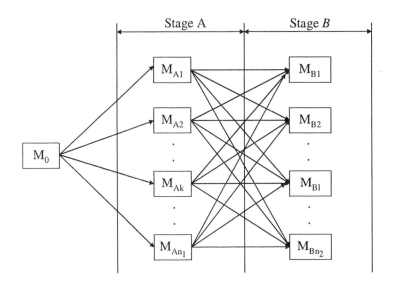

Figure 8-3. Two-stage assembly process

In order to compare H_{A-B} and H_{B-A}, we compute the difference $H_{A-B} - H_{B-A}$

$$H_{A-B} - H_{B-A} = -\frac{1}{V}\frac{T_B}{T}\sum_{k=1}^{n_1} v_{Ak}\, p_{Ak}\, \log_2 p_{Ak} + \frac{1}{V}\frac{T_A}{T}\sum_{k=1}^{n_2} v_{Bk}\, p_{Bk}\, \log_2 p_{Bk}$$

Now, we shall study the function $H_{A-B} - H_{B-A}$ in some particular cases:

- Case 1:

$$\left(v_{Ak}\right)_{k=1,..,n_1} = V_A,\ \left(v_{Bk}\right)_{k=1,..,n_2} = V_B,\ \left(p_{Ak}\right)_{k=1,..,n_1} = P_A = \frac{1}{n_1},$$

$$\left(p_{Bk}\right)_{k=1,..,n_2} = P_B = \frac{1}{n_2}$$

This case corresponds to equal added values and equal selection probabilities of module variants at the same assembly stage.

Thus, $H_{A-B} - H_{B-A} = \dfrac{V_A}{V} \dfrac{T_B}{T} \log_2 n_1 - \dfrac{V_B}{V} \dfrac{T_A}{T} \log_2 n_2$

$$H_{A-B} \leq H_{B-A} \Leftrightarrow \frac{\log_2 n_1}{\log_2 n_2} \leq \frac{V_B T_A}{V_A T_B}$$

Without loss of generality,

let $V_B = vV_A$ and $T_B = tT_A$, where $v > 0$ and $t > 0$, thus

$$H_{A-B} \leq H_{B-A} \Leftrightarrow \frac{\log_2 n_1}{\log_2 n_2} \leq \frac{v}{t}$$

Suppose $\dfrac{v}{t} = 1$. For $1 \leq n_1 \leq n_2$, we have $0 \leq \log_2 n_1 \leq \log n_2$. This

gives $\dfrac{\log_2 n_1}{\log_2 n_2} \leq 1 \Leftrightarrow H_{A-B} \leq H_{B-A}$. Subsequently, sequence A-B is preferred to sequence B-A due to lower variety induced complexity. Note that the delayed product differentiation principle also suggests placing A prior to B. In effect, this principle would provide similar results to those suggested by the minimum variety induced complexity principle if the selection probabilities of module variants at each assembly stage are the same and the rates at which value is added in the course of time at each process are equal. However, note that in the case when $\dfrac{v}{t} \neq 1$, sequence A-B is preferred to sequence B-A if and only if $n_1 \leq (n_2)^{v/t}$. For example, if $\dfrac{v}{t} = 1.5$ and $n_2 = 3$ then sequence A-B should be chosen if $n_1 \leq 3^{1.5} = 5.196 \Rightarrow n_1 \in \{1,2,3,4,5\}$. If $n_1 \geq 6$, then sequence B-A has a lower variety induced complexity.

- <u>Case 2</u>: $(v_{Ak})_{k=1,..,n_1} = V_A$, $(v_{Bk})_{k=1,..,n_2} = V_B$

In order to be able to study the effects of the selection probabilities on the assembly sequence, we suppose that the module variants assembled at one stage have the same added values. Thus, we obtain

$$H_{A-B} - H_{B-A} = -\frac{V_A}{V}\frac{T_B}{T}\sum_{k=1}^{n_1} p_{Ak} \log_2 p_{Ak} + \frac{V_B}{V}\frac{T_A}{T}\sum_{k=1}^{n_2} p_{Bk} \log_2 p_{Bk}$$

$$H_{A-B} - H_{B-A} \leq 0 \Leftrightarrow \frac{\displaystyle\sum_{k=1}^{n_1} p_{Ak} \log_2 p_{Ak}}{\displaystyle\sum_{k=1}^{n_2} p_{Bk} \log_2 p_{Bk}} \leq \frac{V_B T_A}{V_A T_B} = \frac{v}{t}$$

This means that sequence *A-B* is preferred to sequence *B-A* if and only if the quotient of the Shannon entropies is less than the quotient of the rate by which value is added at stage *B* over the rate by which value is added at stage *A*.

In order to determine the optimal assembly sequence if $\dfrac{v}{t} = 1$, it is sufficient to compare both Shannon entropies $H_A = -\displaystyle\sum_{k=1}^{n_1} p_{Ak} \log_2 p_{Ak}$

and $H_B = -\displaystyle\sum_{k=1}^{n_2} p_{Bk} \log_2 p_{Bk}$

Now, in order to gain more insights, suppose that at each stage, only two module variants are assembled, which means that $n_1 = n_2 = 2$. Thus,

$$H_{A-B} - H_{B-A} \leq 0 \Leftrightarrow$$

$$f(p_{A1}, p_{A2}, \frac{v}{t}) = H(p_{A1}, 1 - p_{A1}) - \frac{v}{t} H(p_{B1}, 1 - p_{B1}) \leq 0,$$

where

$$H(p_{A1}, 1 - p_{A1}) = -p_{A1} \log_2 p_{A1} - (1 - p_{A1}) \log_2 (1 - p_{A1}), \text{ and}$$
$$H(p_{B1}, 1 - p_{B1}) = -p_{B1} \log_2 p_{B1} - (1 - p_{B1}) \log_2 (1 - p_{B2})$$

Figure 8-4 depicts the binary Shannon entropy[3] weighted by different values of v/t. One can notice that if $p_{A1} = p_{B1}$ (subsequently $p_{A2} = p_{B2}$), then the value of (v/t) determines the configuration of the assembly sequence. In effect, if $(v/t) > 1$, then placing assembly stage A first will result in lower complexity. However, if $(v/t) < 1$, then stage B should be placed prior to A. In the case when $p_{A1} \neq p_{B1}$, the assembly sequence depends on the values of each variable, namely p_{A1}, p_{A2} and $\dfrac{v}{t}$.

Solving the equation $\dfrac{v}{t} H(p_{B1}, 1 - p_{B1}) = 1$ provides two solutions p_{B1}^1 and p_{B1}^2. In effect, if $(v/t) = 1.15$, $p_{B1}^1 \approx 0{,}3$ and $p_{B1}^2 \approx 0{,}7$. It follows that for $p_{B1} \in [0.3, 0.7]$, we have $\dfrac{v}{t} H(p_{B1}, 1 - p_{B1}) > 1$ (see figure 8-4). Since $\forall p_{A1} \in [0,1]$ the values that are taken by the binary Shannon entropy $H(p_{A1}, 1 - p_{A1})$ are usually less than or equal to 1 $H(p_{A1}, 1 - p_{A1}) - \dfrac{v}{t} H(p_{B1}, 1 - p_{B1}) < 0$. Note that when $\dfrac{v}{t} \to \infty$ (This corresponds to the case when the pace at which value is built up at stage B is very high), $p_{B1}^1 \to 0$ and $p_{B1}^2 \to 1$, in other words $\forall p_{A1} \in [0,1]$ we will have $H_{A-B} - H_{B-A} \leq 0$. It follows that if the value added at stage B is very high, stage A should usually be placed before stage B regardless of which selection probabilities of module variants are involved.

In the case when $p_{A1} \in \left]0, p_{B1}^1\right[$ and $p_{A1} \in \left]p_{B1}^2, 1\right[$, the results cannot be generalized and the decision about the assembly process configuration depends on the values taken by each variable.

---

[3]   The function $H(p, 1-p)$ is called binary Shannon entropy since it is computed on the basis of two values p and $(1-p)$.

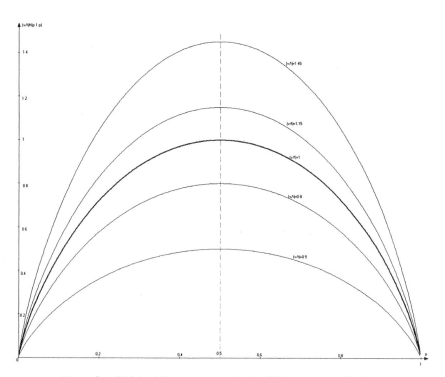

Figure 8-4. Weighted Shannon entropies for different values of (v/t)

- Case 3: General case:

In the general case, stage assembly *A* should be placed before assembly stage *B* if and only if

$$H_{A-B} - H_{B-A} \leq 0 \Leftrightarrow \frac{\sum\limits_{k=1}^{n_1} v_{Ak} \, p_{Ak} \, \log_2 p_{Ak}}{\sum\limits_{k=1}^{n_2} v_{Bk} \, p_{Bk} \, \log_2 p_{Bk}} \leq \frac{1}{t}$$

The study of this inequality when the selection probabilities and added values are arbitrary is quite difficult. In order to gain more insights and to show the utility of the principle of minimum variety induced complexity, we suppose again that $n_1 = n_2 = 2$. In addition, let $v_{A2} = \alpha v_{A1}$ and $v_{B2} = \beta v_{B1}$. Subsequently, $H_{A-B} - H_{B-A} \leq 0 \Leftrightarrow$

$$-p_{A1} \log_2 p_{A1} - \alpha(1 - p_{A1}) \log_2(1 - p_{A1}) \leq$$
$$-\frac{1}{t} \frac{v_{B1}}{v_{A1}} \{ p_{B1} \log_2 p_{B1} + \beta(1 - p_{B1}) \log_2(1 - p_{B1}) \}$$

To illustrate the effects of assembling module variants with very different added values at one stage, we ascertain the selection probabilities. $p_{A1} = p_{B1} = 1/2$. From this, it follows that the condition to be satisfied in order to justify placing stage A before B is: $\dfrac{1 + \alpha}{1 + \beta} \leq \dfrac{1}{t} \dfrac{v_{B1}}{v_{A1}}$. Consequently, the presence of cost intensive module variants drives the placement of the corresponding stage to the end of the assembly process.

Now, we will examine the effects of commonality on the variety induced complexity in the assembly process. Commonality refers to the multiple uses of a few module variants across several product variations. When $n_1 = n_2 = 2$, the substitution of both module variants at one stage by a single module increases commonality. We shall therefore study the impacts of commonality on the variety induced complexity.

Suppose that through a redesign of the product, we replace both module variants M_{A1} and M_{A2} that are assembled at stage A by a single module M_A. In this case, a functional congestion of the new module is necessary since it should perform both functions of M_{A1} and M_{A2}. Therefore, it can generally be assumed that $v_A > v_{A1}$ and $v_A > v_{A2}$.

However, $v_A < v_{A1} + v_{A2}$ because the substitution of two module variants by a common module avoids the duplication of components or interfaces. Consequently, the variety induced complexity that is triggered at stage A is reduced to 0 since the probability of selection of module M_A is equal to 1. Thus variety induced complexity in the assembly process is equal to the complexity brought about by the second stage and subsequently

$$(H_{A-B})_1 = -\frac{T_B}{VT}(v_{B1} p_{B1} \log_2 p_{B1} + v_{B2} p_{B2} \log_2 p_{B2}).$$

Note that the entropy measure does not capture the increase of added value of the common module. In effect, the weighted Shannon entropy introduced at section 4.2 only measures complexity that is triggered by *variety*.

On the other hand, suppose commonality is introduced at stage B so that M_{B1} and M_{B2} are both replaced by module M_B with $v_B > v_{B1}$, $v_B > v_{B2}$ and $v_B < v_{B1} + v_{B2}$. Thus, the total variety induced complexity is $(H_{A-B})_2 = -\dfrac{1}{V}(v_{A1}p_{A1}\log_2 p_{A1} + v_{A2}p_{A2}\log_2 p_{A2})$. It is clear that when there are no assembly sequence constraints, it is more adequate to place the assembly of the common component at the first stage. This also corresponds to the results that would be suggested by the delayed product differentiation principle. However, the minimum variety induced complexity principle provides an additional result. In effect, it suggests introducing the common module at the stage with higher weighted Shannon complexity. This way, the total variety induced complexity can be minimized.

Now consider the case when assembly sequence constraints oblige placing stage A prior to stage B. Furthermore, it might be necessary to make a choice concerning the stage at which the common module should be introduced due e.g. to design team capacity constraints. The delayed product differentiation principle would propose to introduce commonality at stage A. But the variety induced complexity principle suggests comparing both quantities $(H_{A-B})_1$ and $(H_{A-B})_2$. This provides

$$(H_{A-B})_1 - (H_{A-B})_2 = -\frac{T_B}{VT}(v_{B1}p_{B1}\log_2 p_{B1} + v_{B2}p_{B2}\log_2 p_{B2})$$

$$+\frac{1}{V}(v_{A1}p_{A1}\log_2 p_{A1} + v_{A2}p_{A2}\log_2 p_{A2})$$

Thus, if $(H_{A-B})_1 - (H_{A-B})_2 < 0$, it is more adequate to introduce the common module at stage A. However, if $(H_{A-B})_1 - (H_{A-B})_2 > 0$, then it is better to place commonality at the next stage. Note that in the case when $(H_{A-B})_1 - (H_{A-B})_2 = 0$, the delayed product differentiation principle should be applied, thereby resulting in the placement of stage A first.

5.   SUMMARY AND CONCLUSIONS

In this paper, we have presented the insufficiencies of the delayed product differentiation principle. By means of a simple example from the computer industry in which the degree of product modularity is very high, we have demonstrated that this principle cannot support optimal decisions concerning how variety should proliferate throughout the assembly process. Furthermore, we have dealt with the potential problems that may be triggered by the application of this principle in assemble-to-order environments. To fill this gap, the minimum variety induced complexity principle is introduced. It is a complement to the first principle and builds upon the weighed Shannon entropy. The proposed measure evaluates the complexity due to the proliferation of product variety throughout the assembly process. The variety induced complexity depends on three main variables, namely the selection probabilities, value added and assembly time of each module variant.

The results that are attained during the discussion of the two stage assembly process can be generalized for an *n*-stage process in the following way:

- If the selection probabilities of module variants in each stage are equal and the pace at which value is added throughout the assembly process is fairly constant, then both principles would lead to the same result. Therefore, it is adequate to delay the proliferation of variety toward the end of the process.

- If an assembly stage involves an exponential increase of the rate at which value is added, this stage should be placed at the end of the process regardless of the selection probabilities.

- In an assemble-to-order environment, if the selection probabilities of the module variants are very different and the paces at which value is added are very variable, it is necessary to configure the assembly process in such a way that the total value of complexity is kept at a minimum level.

- It is more advantageous to assemble the common module at the beginning of the process than at a subsequent stage. In so doing, the common module can be considered as a part of the basic component (product platform), thereby triggering no extra variety induced complexity. Note that though the placement of a common module somewhere in the middle of the process would generate no direct complexity (Entropy at that stage is equal to 0), it generates an indirect complexity. This is because the proposed entropy measure is a function that increases in the assembly lead time. The higher the number of assembly stages after the first variety proliferation, the higher the variety induced complexity.

- The decision about which module variants should be eventually substituted by a common module can be supported by the entropy measure. The alternative that strongly decreases complexity has to be chosen.

An implicit assumption of our model is that the selection probabilities are independent, which means that the selection of a module variant at one stage does not influence the selection probabilities at subsequent stages. Therefore, this work can be extended by relaxing this assumption. In order to achieve this goal, we have to consider the Shannon entropy defined for conditional probabilities. This way, we can examine the effect of the so-called "blocking" (Maroni 2001) in order to reduce variety induced complexity. Blocking refers to a variety steering action that restrains the mixing and matching possibilities of module variants. It can be described by the following rule: "If module variant 1 is selected, then select module variant 2". In terms of probabilities, this would mean that the conditional selection probability of module variant 2 knowing that module variant 1 has already been chosen is equal to one.

Furthermore, the proposed model assumes must-modules at each assembly stage. In other words, each module family must be represented by one module variant in each product variation. Subsequently, the model does not consider the impacts of options (can-modules) on the variety induced complexity. Each assembly stage in which option variants are assembled would involve two distinct probabilities. The first probability is about the event whether options from that assembly stage would be selected at all. The second one is the conditional probability that an option variant would be chosen knowing that the first event has occurred. This extension will enable us to study the complexity effects of options and to quantitatively measure the advantage of some variety steering actions, e.g. the packaging of options.

REFERENCES

Alderson, Wroe (1950): Marketing efficiency and the principle of postponement, *Cost and Profit Outlook*, Vol. 3, pp. 15-18.

Ashby, Ross, W. (1957): *An Introduction to cybernetics*, 2nd Edition, London: Chapman & Hall LTD 1957.

Baldwin, Carliss Y. / Clark, Kim B. (2003a): Managing in an Age of Modularity, in: Raghu Garud / Arun Kumaraswamy / Richard N. Langlois (Eds.): *Managing in the Modular Age – Architectures, Networks, and Organizations*, Malden et al.: Blackwell Publishing 2003, pp. 149-171.

Bertalanffy, Ludwig V. (1976): *General Systems Theory*, Revised Edition, New York: George Braziller 1976.

Blecker, Thorsten / Friedrich, Gerhard / Kaluza, Bernd / Abdelkafi, Nizar / Kreutler, Gerold (2005): *Information and Management Systems for Product Customization*, Boston et al.: Springer, 2005.

Bucklin, Louis P. (1965): Postponement, speculation and the Structure of Distribution Channels, *Journal of marketing research*, Vol. 2, No. 1, pp. 26-31.

Christopher, Martin (2005): *Logistics and Supply Chain Management – Creating Value-Adding Networks*, 3rd Edition, Harlow et al: Prentice Hall 2005.

Duray, Rebecca / Ward, Peter T. / Milligan, Glenn W. / Berry, William L. (2000): Approaches to mass customization: configurations and empirical validation, *Journal of Operations Management*, Vol. 18, No. 6, pp. 605-625.

Ericsson, Anna / Erixon, Gunnar (1999): *Controlling Design Variants: Modular Product Platforms*, Dearborn / Michigan: Society of Manufacturing Engineers 1999.

Feynman, Richard P. (1991): *Vorlesungen über Physik*, vol. 1, München / Wien: Oldenburg 1991.

Garud, Raghu / Kumaraswamy, Arun (2003): Technological and Organizational Designs for Realizing Economies of Substitution, in: Raghu Garud / Arun Kumaraswamy / Richard N. Langlois (Eds.): *Managing in the Modular Age – Architectures, Networks, and Organizations*, Malden et al.: Blackwell Publishing 2003, pp. 45-77.

Klir, George J. / Folger, Tina A. (1988): *Fuzzy Sets, Uncertainty, and Information*, New Jersey: Prentice Hall, Englewood Cliffs 1988.

Lampel, Joseph / Mintzberg, Henry (1996): Customizing Customization, *Sloan Management Review*, Vol. 38, No. 1, pp. 21-30.

Lee, Hau L. (1996): Effective Inventory and Service Management through Product and Process Redesign, *Operations Research*, Vol. 44, No. 1, pp. 151-159.

Lee, Hau, L. / Tang, Christopher S. (1997): Modelling the Costs and Benefits of Delayed product Differentiation, *Management Science*, Vol. 43, No. 1, pp. 40-53

Maroni, Dirk (2001): *Produktionsplanung und -steuerung bei Variantenfertigung*, Frankfurt am Main et al.: Peter Lang 2001.

Martin, Mark V. / Ishii, Kosuke (1996): Design For Variety: A Methodology For Understanding the Costs of Product Proliferation, *Proceedings of The 1996 ASME Design Engineering Technical Conferences and Computers in Engineering Conference*, California, August 18-22, 1996, URL: http://mml.stanford.edu/Research/Papers/1996/1996.ASME.DTM.Martin/1996.ASME.DTM.Martin.pdf (Retrieval: April 01, 2005).

Martin, Mark V. / Ishii, Kosuke (1997): Design For Variety: Development of Complexity Indices and Design Charts, *Proceedings of DETC'97, 1997 ASME Design Engineering Technical Conferences*, Sacramento, September 14-17, 1997, URL: http://mml.stanford.edu/Research/Papers/1997/1997.ASME.DFM.Martin/1997.ASME.DFM.Martin.pdf (Retrieval: April 01, 2005).

Pine II, B. Joseph (1993): *Mass Customization: The New Frontier in Business Competition*, Boston, Massachusetts: Harvard Business School Press 1993.

Schilling, Melissa A. (2003): Toward a General Modular Systems Theory and its Application to Interfirm Product Modularity, in: Raghu Garud / Arun Kumaraswamy / Richard N. Langlois (Eds.): *Managing in the Modular Age – Architectures, Networks, and Organizations*, Malden et al.: Blackwell Publishing 2003, pp. 172-214.

Shannon, Claude E. (1948): A Mathematical Theory of Communication, *The Bell System Technical Journal*, Vol. 27, pp. 379-423.

Sharman, G. (1984): The rediscovery of logistics. *Harvard Business Review*, Vol. 62, No. 5, pp. 119-126.

Sivadasan, Suja / Efstathiou, Janet / Calinescu, Ani / Huaccho Huatuco, Luisa.: Advances on measuring the operational complexity of supplier-customer systems, *European Journal of Operational Research* (Article in Press).

Ulrich, Karl (1995): The role of product architecture in the manufacturing firm, *Research Policy*, Vol. 24, No. 3, pp. 419-440.

Van Hoek, Remko I. (1997): Postponed manufacturing: a case study in the food supply chain, *Supply Chain Management*, Vol. 2, No. 2, pp. 63-75.

Van Hoek, Remko I. (2001): The rediscovery of postponement a literature review and directions for research, *Journal of Operations Management*, Vol. 19, No. 2, pp. 161-184.

Zinn, Walter / Bowersox, Donald, J. (1988): Planning physical distribution with the principle of postponement, *Journal of Business Logistics*, Vol. 9, No. 2, pp.117-136.

Chapter 9

A NEW MIXED-MODEL ASSEMBLY LINE PLANNING APPROACH FOR AN EFFICIENT VARIETY STEERING INTEGRATION

Stefan Bock

University of Paderborn, Graduate School Dynamic Intelligent Systems

Abstract: In order to deal with the extreme complexity occurring within mass customization production processes today, appropriate variety steering and formation concepts are mandatory. Those are responsible for a customer oriented variant definition which simultaneously reduces internal complexity. In order to achieve mass production at least costs, assembly lines are still attractive means. By avoiding transportation and storage as well as, in particular, by specifically training the employed workers, assembly lines yield substantial reductions of variable unit costs. However, by producing a mass customization variant program with billions of different constellations on the same line, an oscillating capacity use can be observed. Obviously, planning the structure of the used assembly lines and variety steering are strongly interdependent decision problems whose coping is decisive for efficient mass customization. Unfortunately, an integration of both decision levels currently fails because of lacking adequate approaches for mixed-model assembly line balancing. Since known concepts are still based on integral product architectures, they neither correspond to existing steering approaches nor do they cope with the extreme complexity of mass customization processes. Consequently, the present paper sketches a new balancing approach with a modular variant definition. In addition to this, the new model comprises a sophisticated personnel planning. In order to determine systematically appropriate line layouts, a randomized parallel Tabu Search algorithm is generated and analyzed. This approach was designed for the use in an ordinary LAN of personal computers which can be found in almost all companies today. In order to validate its utilizability in companywide networks, results measured for constellations with oscillating background loads are presented.

Key words: Variety Steering and Formation, Mass Customization, Mixed-model assembly line balancing, Distributed algorithms

1.   INTRODUCTION

In order to handle extreme complexity occurring for Mass Customization production processes, the application of appropriate variety formation and steering approaches is of significant importance. Those concepts try to determine variant programs, which enable, on the one side, comprehensive satisfaction of customer needs and simultaneously keep the internal complexity costs as low as possible. Most of them involve a modular variant definition which decides about offered product modules whose combinations are frequently grouped in platforms. On account of these definitions, the variant program arises by module combinations within the manufacturing process.

Despite their inflexible structure, assembly lines are still attractive means of an efficient execution of mass production processes (Becker and Scholl (2005)). By avoiding unnecessary transportation, storage and in particular by specifically training the employed workers, the use of these mass production systems leads to a significant reduction of variable costs. Originally, the use of these systems was restricted to single product scenarios, where large-scale productions of homogeneous products are realized. Consequently, a stationary capacity demand occurs throughout its use. In contrast to this, mixed-model assembly lines have to handle oscillating demands. By differing from each other in terms of specific features or technical requirements, the production of different variants demands varying tasks to be executed at the line. Therefore, in times of Just-in-Time philosophies, the line has to be flexible enough to execute almost all possible sequences of product types without causing work overload. In order to achieve this, the determination of the line layout has to anticipate and rate possible scenarios potentially occurring during the production processes. However, by switching to Mass Customization, industrial assembly lines frequently have to manufacture theoretical production programs comprising more than one billion of different variants. Consequently, an integrated handling of variety steering and assembly line balancing seems to be promising. However, owing to the necessary modular program determination within variety steering concepts, known mixed-model assembly line balancing approaches do not correspond with this variant architecture common in Mass Customization (see Blecker et al. (2004) p.233). In contrast to variety formation and variety steering concepts, assembly line balancing approaches are using an integral architecture. Here, each possible variant is interpreted as an additional complete product type. Thus, the link necessary for an appropriate interaction between these two interdependent decision levels is still lacking.

Furthermore, owing to their integral variant determination, known assembly line balancing approaches cannot cope with the extreme complexity of Mass Customization production processes. For instance, these approaches frequently use an objective function which generates a specific contribution for each possible theoretical variant. Consequently, programs comprising more than a billion variants cannot be handled anymore. In addition, a reasonable information management cannot be applied within the balancing approach. By treating a variant as a complete product, only two information scenarios are assumed in literature. At first, former approaches (Thomopoulos (1970)) assume that the occurrence probability of each possible variant is known in advance. Consequently, variant-dependent frequencies were presupposed and used. As stated by many authors, the availability of such a detailed knowledge seems to be very unrealistic (Scholl (1999) pp.87-94). Consequently, in contrast to this, Thomopoulos himself and many other authors proposed additional concepts being independent of an expected model mix (Scholl (1999) (chapter 3.2.2.3), Becker and Scholl (2005) p.15, Bukchin et al. (2002)). However, both kinds of approaches seem to be unreasonable. Even though it is frequently not possible to accurately estimate the occurrence probability of entire variants, this can absolutely be the case for specific product features. More precisely, many automotive manufacturers cannot estimate the frequencies in which a specific variant of a new car model will be produced within the considered planning horizon. However, in contrast to this, information regarding the number of sport transmissions possibly installed can be determined by market surveys or analyses of customer behavior recently observed. But to integrate this available information in a decision support system for assembly line balancing, a problem definition has to be derived which uses a corresponding modular variant definition.

Altogether, it can be stated that in order to deal with the existing shortcomings, it is mandatory to develop a planning approach for mixed-model assembly lines which is based on a modular variant definition. This is provided by the present paper.

2.   VARIETY STEERING METHODS

As mentioned above, the complexity of modern mass production processes is significantly driven by the complexity of the offered variant program. In order to provide suitable variety programs, specific variant management approaches are developed in literature. Specifically, variety formation and variety steering concepts can be distinguished (see Blecker et al. (2004) p.233). The first ones aim to reduce customer's complexity (i.e. the so-called external complexity) in order to efficiently find appropriate variants from the entire product assortment. This selection process should directly address customer's specific needs and requirements. It is based on the simple obtained cognition that customers are merely interested in the choice that just ensures their individual demands. Specifically, a larger variety is dispensable and therefore not appreciated by the customers. In contrast to variety formation, variety steering concepts additionally address internal complexity issues. Here, it is intended to determine a variety program that meets the customer requirements but at the same time keeps the internal complexity costs as low as possible (Blecker et al. (2003)).

In order to deal with this tradeoff, variety steering approaches frequently focus on modularity. Specifically, a modular product definition is used (see Duray et al. (2000) p.608). In the respective literature, modularity is introduced and defined in various facings. But all of them characterize a manufactured product by specific features or options. In order to illustrate this and show the proliferation of the resulting theoretical production program, the example of Rosenberg (1996) is frequently cited. It demonstrates the basic variant structure of an ordinary middle-class car offered by a German automotive manufacturer, which is depicted in the following table. As illustrated there, this simple modular variant structure with only 9 obligatory features and 14 complementary ones already results in a theoretical variant program comprising in total 8,918,138,880 variants. Note that these variants can be chosen by customers and have to be produced on the same line. In times of Mass Customization, such a large total number of variants is not unusual. For instance, Piller et al. (2003) reports from a well-known mass customizer of sport shoes who offers approximately $3 \cdot 10^{21}$ variants over the Internet.

On account of the modular variant definition, in the final assembly process, end products arise by the combination of selectable product features belonging to the respective orders. By varying these attribute values, an oscillating demand occurs in the assembly line production process. Therefore, the complexity of the variety program significantly affects the resulting manufacturing costs. Unfortunately, these dependencies cannot be appropriately detected without a comprehensive understanding of the final production process. On the other hand, these data are necessary prerequisites for appropriate variety steering decisions.

Table 9-1. Variant structure of a German automotive manufacturer for a middle-class car.

Feature	Description	Obligatory feature? (yes/no)	Number of values	Total number of theoretical variants
1	Engines	Yes	7	7
2	Transmissions	Yes	3	7·3=21
3	Brake systems	Yes	2	21·2=42
4	Car body variants	Yes	2	42·2=84
5	Chassis frames	Yes	2	84·2=168
6	Colors	Yes	15	168·15=2520
7	Seat coverings	Yes	8	2520·8=20160
8	Type of glazing	Yes	2	20160·2=40320
9	Window lift	Yes	2	40320·2=80640
10	Front spoiler	No	1	80640·(1+1) =161280
11	Rear end spoiler	No	1	161280·2=322560
12	Fog lamp versions	No	1	322560·2=645120
13	Type of rev meter	No	1	645120·2 =1290240
14	Multi-functional display	No	1	1290240·2 =2580480
15	Radio types	No	3	2580480·4 =10321920
16	Wing mirror (co-driver side)	No	2	10321920·3 =30965760
17	Type of sunroof	No	2	30965760·3 =92897280
18	Central locking system	No	1	92897280·2 =185794560
19	Type of trims	No	1	185794560·2 =371589120
20	Type of antenna	No	2	371589120·3 =1114767360
21	Air conditioning	No	1	1114767360·2 =2229534720
22	Seat heating	No	1	2229534720·2 =4459069440
23	Airbag	No	1	4459069440·2 =8918138880

Consequently, by deciding about the composition of the production program, the variety steering process requires substantial information about the layout of the assembly line and its usage. Thus, in order to provide decision support for dealing with these important interdependencies between the variety steering process and the assembly line balancing level, proposed sophisticated hierarchical planning approaches for assembly line balancing and sequencing (see Scholl (1999) pp.106-112 or Domschke et al. (1996) pp.1487/1488) have to be extended. More precisely, an integration of the variety steering level seems to be mandatory. By doing so, decisions for integrating, retaining or eliminating specific sets of variant attributes or features can be supported by information originating from the subsequent balancing level. Here, consequences of a modified variant complexity within the offered program are anticipated through illustrating the estimated resulting cost differences and detailed information about the changed line layout. This supports a considerably better understanding of internal complexity dependencies according to the incurred manufacturing costs.

However, this promising integration of variety steering and assembly line layout planning necessarily requires an appropriate approach for mixed-model assembly line balancing. Unfortunately, known approaches for mixed-model assembly line balancing completely neglect variety steering aspects. For instance, to the knowledge of the author, there is currently no approach proposed in literature that comprises a modular variant definition. By defining a production program only as a set of complete products, these approaches are not compatible with variety steering processes. Thus, the present paper sketches a new balancing approach which is based on a modular variant determination. But before this balancing approach (problem model and solution approach) is depicted in the Sections 4 and 5, a brief overview of balancing approaches currently proposed in literature is given.

3.   KNOWN BALANCING APPROACHES IN LITERATURE

By analyzing the respective literature, it becomes visible that balancing mixed-model assembly lines has been a vital area of research for many decades. Here, depending on the number of product types to be produced on the line, single-model problems are distinguished from mixed-model approaches. Various single-model approaches are defined as specific versions of the well-known Simple Assembly Line Balancing Problem (SALBP), in which the layout planning process is reduced to a simple task-station assignment (see among others Scholl and Becker (2005), Scholl (1999), Pinnoi and Wilhelm (1997), and the literature cited there). Owing to their restrictive requirements, SALBP-based models cannot appropriately map realistic balancing problems occurring for the use of modern assembly lines. Consequently, many extensions of this basic model are proposed in literature integrating parallel stations, cost aspects, process alternatives or intended job enrichment (see Becker and Scholl (2005), Bukchin and Masin (2004), Rosenberg and Ziegler (1992), Amen (2005), Pinnoi and Wilhelm (1997), Pinto et al. (1983)).

However, in times of keen competition resulting in customer orientation and Mass Customization policies, mixed-model formulations increasingly gain attention. By the simultaneous consideration of several variants, balancing mixed-model assembly lines is much more difficult than the single-model case. Owing to the arising oscillating variant-dependent capacity demand, a task allocation additionally has to smooth the resulting execution times in the different stations for the variant program to be produced. Otherwise, inefficient work overload or idle times are likely to happen during the production process (Becker and Scholl (2005) p.15).

Consequently, sophisticated objective functions have to be designed to incorporate those important aspects. For instance, Thomopoulos (1970) proposes minimizing the total sum of absolute differences between all occurring station times of the different variants and the average station time. This performance measure is denoted as the smoothed station objective (Bukchin et al. (2002) p.411). In contrast to this, the approach of Macaskill (1972) intends to minimize the occurring idle times within the production process. Fremerey (1991) introduces a station coefficient of variation measure while the paper of Bukchin (1998) provides the model variability and the bottleneck measure. Additionally, Bukchin shows by computational simulations that his bottleneck measure outperforms the others in showing a significant correlation with the operational throughput objective (Bukchin (1998) pp. 2675-2684 and Bukchin et al. (2002) p.411).

However, by being based on a pure integral variant definition, known mixed-model assembly line balancing approaches are neither corresponding to variety steering approaches nor can they deal with an extremely large complexity typical for Mass Customization manufacturing. Consequently, the following subsection provides a new balancing model which is based on a modular variant architecture. Note that this kind of architecture directly corresponds to variety steering approaches.

4.   THE BALANCING MODEL

In this section, a new planning model is introduced. Before Subsection 4.2 provides the corresponding mathematical definition, a brief description of the modular variant architecture is given.

4.1   Modular variant definition

As already mentioned above, traditional balancing approaches make use of an integral variant definition. Consequently, an integration of variety steering decision processes cannot be provided. In addition, integral variant definitions are not useful for Mass Customization production processes. By treating and numbering each possible variant as an additional product, the use of an integral variant definition results in very complex but at the same time unnecessary calculations. Specifically, by analyzing the example of Rosenberg (1996), it becomes obvious that by assuming an average number of two assigned tasks per station, for the objective function calculation, there are 17,836,277,760 task-variant constellations to be respected per station. Note that for each of them, a specific contribution to the objective function has to be examined.

In contrast to this, in the modular definition of the new balancing approach, the variant program arises by the definition of the product features relevant to the customers. For each characterizing feature the offered values have to be defined, respectively. In addition to this, for each task to be executed for some variant at the line the product attribute has to be examined which influences its implementation. Such a feature is denoted as the relevant feature of the respective task. Note that it is assumed that for each task exactly one relevant feature exists. This must be guaranteed by the task and feature definition phase. However, if a situation occurs where the execution of a specific task depends on more than a single variant feature there are two measures thinkable. First, the task can be split into two or more sub-tasks. Second, if a task split is not possible, the relevant features have to be combined into a single one. This aggregated feature then comprises all existing value combinations of the original ones reduced by the infeasible possibilities. Consequently, by giving each task its relevant feature, task execution times are now defined according to all possible values of the respective relevant feature.

What effects does this modification have for the example introduced above? First of all, the modified approach uses now a variant definition which is directly compatible with variety steering and formation approaches. This enables an integrated determination of the offered variant program. By providing a more accurate estimation of the internal complexity costs occurring within the manufacturing processes, decisions for installing, retaining or eliminating product modules or features can be done more sophisticatedly. A second, not less important, effect of the modified modular variant definition is the achieved complexity reduction within the problem model. For its understanding, the simple example of Rosenberg (1996) is examined once more. Let us additionally assume that there are at most 3 tasks allocated per station. In addition to this, in the worst case scenario, these tasks have non-identical relevant features. Finally, in this worst case constellation, these features possess the maximum number of selectable values, respectively. Consequently, for the example of Rosenberg, in the new model, this scenario would comprise the examination of only $15 \cdot 8 \cdot 7 = 840$ constellations in this specific station. Note that for at most three tasks per station, the average number of constellations to be considered per station would be significantly lower.

Another important aspect of assembly line balancing, considered in the sketched model, deals with the employments of workers and floaters. In particular, an adequate estimation of the size of the floater pool is of significant importance. Note that on account of their engagement in different stations, floaters receive a significantly higher payment than ordinary workers, who always perform identical tasks at the same station. Consequently, by the determination of the size of the floater pool, considerable proportions of the available capital resources are already fixed. In order to deploy floaters more efficiently, it is assumed that they additionally work aside the line in a specific closely located offline area. Specifically, if a floater is currently not engaged at the line, he or she performs non-time critical tasks aside it. But, if work overload occurs, i.e. if the number of workers currently employed at some station is not sufficient to complete the processing of some variant in the remaining cycle time, floaters are assigned to the respective station. This, now reduced, processing time is considered as the employment duration of the allocated floaters. Note that the line layout structure must guarantee that a floater employment can ensure a processing within the predetermined cycle time for each variant.

4.2   Mathematical definition

This subsection provides a sketch of the entire mathematical definition of the problem formulation. Note that the entire model comprises several additional instruments of industrial mixed-model assembly lines dealing, for instance, with alternative task processing, parallel stations, and distance restrictions. However, in order to simplify the following definitions, these aspects are omitted.

4.2.1   Parameters

The following parameters characterize an instance of the mathematical assembly line balancing problem.

- N: Number of tasks
- F: Number of features
- T: Number of periods belonging to the planning horizon in question
- P: Number of product units to be produced in each period of the planning horizon
- C: Cycle time of the assembly line
- M: Maximum number of stations installable at the line
- S: Necessary investments in period 0 for each station to be additionally installed at the line
- $FV_f (1 \le f \le F)$: Number of possible constellations for the f-th feature
- $P_{f,v} (1 \le f \le F; 1 \le v \le FV_f)$: Assumed frequency in which the v-th constellation of the f-th feature is produced in each period of the entire planning horizon in question. If this information is not available, one can assume an equipartition of all values. Consequently, it holds:

$$\forall f \in \{1,...,F\}: \sum_{v=1}^{FV_f} P_{f,v} = P$$

- MW: Maximum number of workers assignable to a station
- i: Interest rate. This rate gives the average profit rate per period yielded by the company in question.
- $T_i \in \{1,...,F\} (1 \le i \le N)$: Relevant feature of task i
- $t_{i,v,w} (1 \le i \le N; 1 \le v \le FV_{T_i}; 1 \le w \le MW)$: Execution time of task i with employed w workers implementing the v-th value of the relevant feature T_i. Note that it is assumed that up to the application-dependent threshold MW, an allocation of additional workers may result in a reduction of the execution time of the tasks.
- $\Gamma(i) (1 \le i \le N)$: Set of successors of task i in the combined precedence graph
- WW: Wage for a worker per period
- WOF: Offline wage for a floater per time unit
- WF: Wage received by each employed floater per time unit. In addition, $WOF < WF$ is assumed
- WTF: Working time of each floater per period measured in time units. It holds $WTF = P \cdot C$

4.2.2 Binary auxiliary operations

In order to simplify the coming depictions, two specific abbreviations for binary auxiliary operations are introduced. More precisely, during the following calculations, it is frequently necessary to decide about the proportion of two natural numbers. For this purpose, the binary operations "*GT*=Greater than" as well as "*Eq*=Equal" are introduced.

- Greater than *GT:*

$$\forall x, y \in IN_0 : \underbrace{GT(x, y) = \frac{\max\{x, y\} - y}{\max\{1, x - y\}}}_{GT(x,y)=1 \Leftrightarrow x > y}$$

- Equal *Eq*:

$$\forall x, y \in IN_0 : \underbrace{Eq(x, y) = (1 - GT(x, y)) \cdot (1 - GT(y, x))}_{Eq(x,y)=1 \Leftrightarrow x = y}$$

4.2.3 Variables of the model

The following abbreviations entirely define a chosen line layout. For this purpose, each task has to be unambiguously assigned to a station. In addition, those stations have to be equipped with personnel resources.

- $s_{i,m} (1 \leq i \leq N; 1 \leq m \leq M)$: Station allocation variables. Specifically, this binary variable indicates whether the i-th task is assigned to station m
- $w_m (1 \leq m \leq M)$: Number of workers allocated to station m

4.2.4 Objective function

Owing to its long lasting consequences, installing a new assembly line has to be characterized as an investment. Therefore, a layout found is rated by a payoff-oriented Net Present Value (NPV). In order to accurately evaluate the temporal assignment of the occurring payments, an interest rate i is used which determines the average yield the spent capital attains in the company in question. Owing to the fact that the variable payments of all periods are identical, the calculation can be considerably simplified by the use of a present value annuity factor. Consequently, the objective function value can be determined by

$$Minimize\ NPV = Inv + Pay \cdot \frac{(1+i)^T - 1}{i \cdot (1+i)^T}$$

In this calculation, *Inv* determines the investments necessary for the use of the assembly line and therefore paid in advance of the first period. Specifically, *Inv* can be calculated as follows:

$$Inv = \sum_{m=1}^{M} GT\left(\sum_{i=1}^{N} s_{i,m}, 0\right) \cdot S$$

In addition,

$$Pay = \sum_{i=1}^{M} w_m \cdot WW + NF \cdot WTF \cdot (WF - WOF) + FT \cdot WOF$$

In this computation, *NF* determines the number of employed floaters, i.e. the minimum number of floaters necessary to avoid work overload in a worst case scenario. Such a worst case constellation is characterized by the simultaneous production of the most complex variant in each station. Thus, for each station the respective relevant features of all assigned tasks are set to the values which cause the maximum execution times. For the time units in which a floater is not employed at the line, the assembly system has to account only for the difference between both wages (*WF-WOF*). Otherwise, if a floater is assigned to the line, the full floater wage is allocated. Consequently, the number of time units *FT* in which floaters are employed at the assembly line in each period of the planning horizon must be derived. For this purpose, the set of relevant features is defined for each station.

$$\forall m \in \{1,...,M\}:$$

$$RF_m = \left\{ f \ \middle| \ 1 \le f \le F \wedge GT\left(\sum_{i=1}^{N} (s_{i,m} \cdot Eq(T_i, f)), 0\right) = 1 \right\}$$

Obviously, only features belonging to this set may influence the task execution in the m-th station. In order to simplify the following depictions, these features are renumbered for each station by using the permutation rrf_m:

$$\forall m \in \{1,...,M\}: rrf_m : RF_m \rightarrow \{1,...,|RF_m|\}$$

For $f \in RF_m$, $rrf_m(f)$ refers to the new index of this feature relevant for station m. Using rrf_m, a set of all possible constellations of the relevant features is introduced for each station:

$$\forall m \in \{1,...,M\}:$$
$$RVF_m = \left\{ \left(v_1,...,v_{|RF_m|}\right) \mid \forall k \in \{1,...,|RF_m|\}: 1 \leq v_k \leq FV_{rrf_m^{-1}(k)} \right\}$$

Using this subset and the maximum number of available floaters for the m-th station NF_m, the expected durations with floater employment per period can be derived for the m-th station by:

$$\forall m \in \{1,...,M\}: FT_m$$

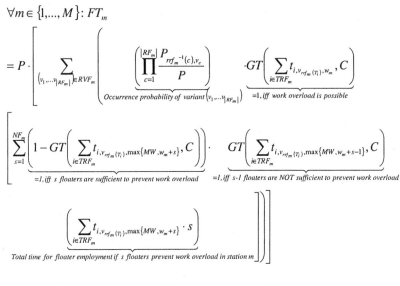

$$using: \forall m \in \{1,...,M\}: TRF_m = \left\{ i \mid 1 \leq i \leq N \wedge s_{i,m} = 1 \right\}$$

Eventually, *FT* determines the total duration of a floater engagement at the line. Therefore, it can be calculated by:

$$FT = \sum_{m=1}^{M} FT_m$$

4.2.5   Restrictions

A generated layout for the mixed-model assembly line is denoted as feasible if it fulfills the following restrictions:

1. Compliance with existing precedence constraints:

$$\forall i \in \{1,...,N\}: \forall j \in \Gamma(i): \sum_{m=1}^{M} s_{i,m} \cdot m \leq \sum_{m=1}^{M} s_{j,m} \cdot m$$

2. Clearly defined task allocation:

$$\forall i \in \{1,...,N\}: \sum_{m=1}^{M} s_{i,m} = 1$$

3. Consistence of task allocation and worker employment. In addition to this, each worker assignment has to respect the predefined upper bound *MW*:

$$\forall m \in \{1,...,M\}: GT\left(\sum_{i=1}^{N} s_{i,m},0\right) \leq w_m \leq MW \cdot GT\left(\sum_{i=1}^{N} s_{i,m},0\right)$$

4. Compliance with the cycle time, i.e. at least NF_m additional floaters are sufficient to prevent work overload in station *m*:

$$\forall m \in \{1,...,M\}:$$

$$\sum_{c \in RF_m} \max\left\{\sum_{i=1}^{N} Eq(T_i,c) \cdot s_{i,m} \cdot t_{i,v,\max\{MW,w_m+NF_m\}} \mid 1 \leq v \leq FV_c\right\} \leq C$$

As described above, for each feature, which is relevant for the station in question, the value that leads to the maximum execution time is chosen.

5. Calculation of the total number of necessary floaters:

$$NF = \sum_{m=1}^{M} NF_m$$

5.   THE DISTRIBUTED TABU SEARCH ALGORITHM

Note that Section 4 provides only a sketch of the complete model definition. Besides the instruments for a modified variant definition and detailed personnel planning described above, the entire model additionally addresses several attributes of modern assembly lines (see Bock (2000) pp.87-145). Therefore, a solution of the entire model comprises in total the following five different classes of variables:

1. **Process alternative selection**: In the model the implementation of each task has to be determined. Thus, different process alternatives can be defined. For each of them the model allows the definition of specific investments, variable costs, processing times, and precedence constraints. In a solution of the model, a subset of variables determines the process alternative to be implemented for each task. By defining necessary investments and consequences of the different process alternatives, the layout to be determined has to identify an efficient compromise between the resulting fixed and variable costs. For instance, while one processing alternative may represent the processing with already existent facilities and therefore comprises no additional investments, another one may propose the procurement of modern facilities resulting in less variable costs and reduced execution times.
2. **Task position at the line**: Each task must be clearly positioned at the line. Therefore, a solution provides for each task a station in which it is executed throughout the planning horizon. However, by introducing distance restrictions between pairs of tasks, it is necessary to provide not only a station assignment but additionally an entire task sequence.
3. **Number of stations**: By allocating tasks and workers as well as floaters to them, stations become installed or not. Thus, a complete layout provides the number of stations at the line. Note that besides the used equipments and facilities, the installation of each station causes additional investments.

4. **Degree of station parallelization**: In order to provide an efficiently balanced layout, it is frequently reasonable to allow layout constellations with parallelized stations.
5. **Personnel employment**: As defined above, the approach provides detailed decision support for personnel planning. Therefore, the number of employed workers is determined for each station. In addition to this, the size of the floater pool is defined by the sum of floater allocations over all stations necessary for an assumed worst case scenario.

In order to solve this complex combinatorial problem, a randomized Tabu Search approach has been designed. By starting with a randomly generated initial layout, this heuristic procedure intends to iteratively improve the solution stored. For this purpose, specific operations which modify the current constellation are applied. Generally, all feasible solutions obtainable by a single application of one of these operations are denoted as the neighborhood of the solution currently stored. The following subsection provides the definition of the neighborhood used throughout the proposed Tabu Search procedure.

5.1   The used neighborhood

By randomly selecting a subset of applicable operations in each move, the generated distributed Tabu Search procedure applies a variable neighborhood throughout the searching process. First, the efficiency of a variable neighborhood was empirically shown for several applications by Hansen and Mladenovic (2001). The entire neighborhood, out of which an equally sized subset is randomly selected in each move, comprises all possible constellations of the following basic operation types:

1. **Swap of two tasks (Type 1)**: In this operation, the station assignment of two tasks is exchanged. Obviously, both tasks must be currently assigned to different stations. However, this operation can be completed only if the intended exchange complies with all existing precedence and distance restrictions, i.e. there must be a feasible position in the task sequence for both exchanged tasks within their new stations.
2. **Task move (Type 2)**: Here, a task is moved to a different station. Again, this operation can be completed only if there is a sequence position for the task within the new station which does not violate any existing restriction.
3. **Increase/Decrease of the number of employed worker (Type 3)**: By applying this operation, modified numbers of assigned workers are tested for the stations at the line. Therefore, for the station in question, a feasi-

ble increase as well as a feasible decrease by one is examined in each move.

4. **Modification of the selected task processing alternatives (Type 4):** This operation examines modified process alternatives for a task currently under consideration.

In order to reduce the complexity of the move generation, in each step of the Tabu Search procedure only a randomly selected subpart of the entire neighborhood is examined to modify the solution currently under consideration. Therefore, at the beginning of each move an initial station *is* is randomly selected. By subtracting and adding predefined parameters *ds* and *de* (with *ds<de*) to *is*, station sets arise which define start (*SS*) and end positions (*ES*) for the operations itemized above. Therefore, it holds:

$$SS = \{s \mid \max\{1, is - ds\} \le s \le \min\{M, is + ds\}\}$$
$$\subseteq ES = \{s \mid \max\{1, is - de\} \le s \le \min\{M, is + de\}\}$$

In each move, every possible operation of the types 1 and 2 which takes at least one of the tasks out of the station set *SS* and the respective destination out of *ES* is examined. In addition, all possible operations of the types 3 and 4 are examined for the tasks currently assigned to the station set *SS*. After the examination of all potential modifications, the best constellation found is executed. Note that this is done independently of the fact whether it leads to an improvement or not. Apart from it, always after a predefined number of unsuccessful moves not improving the best solution currently found, the best modification of the implemented parallelization degree is tested for an existing station. This is done by a possible and feasible unification or separation of neighboring stations.

5.2   The applied dynamic load balancer

In order to accelerate the time-consuming examination processes, all procedures were designed as distributed search algorithms applicable in an ordinary Local Area Network (LAN), connecting modern heterogeneous personal computers. Note that such kind of systems can be found in almost all companies today. However, by using these powerful systems only for ordinary office applications or information transfers, a significant proportion of the entire system performance remains unused today.

Consequently, it seems to be very promising to use these available parts of the computational performance for a more exhaustive examination of the solution space in order to find more suitable layout constellations. Hence, appropriate load balancers have to be designed to share the occurring computational work between the different computers according to their current performance. More precisely, during the computations, this load balancer is accountable for an equal work sharing of the total work in the entire network between all used computers. Note that besides the planning algorithm, the total work comprises all background load applications simultaneously executed in the LAN. Consequently, if this background load changes significantly on some computer, the load balancer has to smooth the total load by balancing the current work distribution of the planning algorithm.

As proposed by Bock and Rosenberg (2000), the Tabu Search algorithm clusters the whole PC-Network of p computers in $O(\sqrt{p})$ teams comprising $O(\sqrt{p})$ members each. After generating a different initial solution by applying a randomized construction procedure, each team independently applies the probabilistic Tabu Search algorithm described above. By dividing the neighborhood examination work among the group members, an accelerated move execution is achieved within each team. Note that this distribution always respects the current performance of each team member. For this purpose, apart from the best found operation, each processor informs the team master (i.e. the lowest numbered node in the team) in every move about its current performance by measuring its examination speed (i.e. number of examined constellations per time unit) in the last move. In the subsequent iteration, the part of the entire neighborhood to be examined in the team in question is distributed according to this information. Since each move takes only about one second, this processing allows a fast work balancing according to a dynamically changing background load caused by additionally executed applications.

6.   COMPUTATIONAL RESULTS

In order to validate its performance and applicability, the outlined algorithm has been tested on different parallel and distributed systems. In the following, this paper provides some results measured in an ordinary SUN Workstation Cluster of the University of Paderborn. Note that the workstations in the network were simultaneously used by students, which causes a dynamically changing background load that is comparable to those occurring in firm networks. Thus, the used test constellation provides a realistic environment for the application of the generated planning procedure. Specifically, the used computers communicate by the use of an ordinary ETHERNET connection. Consequently, since those systems involve large cost factors for information transfer as well as for each message initialization, a challenging benchmark for the use of the proposed algorithm is provided. Altogether, the following analysis consists of the examination of two different test constellations. In a first scenario, the assembly line balancing approach was executed without any additional artificial background load but in parallel to "real applications" launched by students. In order to particularly test the adaptability of the dynamic balancing routine in extreme constellations, a second and even more challenging scenario was initiated. Here, a random load generator was applied. It starts and stops very complex mathematical calculations causing significant load changes at randomly drawn computers of the network. Note that for an efficient use of the computational system, these loads should force the load balancer to execute considerable balancing activities.

Table 9-2 provides the average results obtained for 10 instances in the first scenario. All these tested instances comprised between 60 and 100 tasks each, while the number of features was between 10 and 30 with 1 to 6 selectable values, respectively. By analyzing the measured results, it becomes obvious that despite an existing background load in the network, an almost linear speedup according to the number of executed moves was attained.

Table 9-2. Measured average results for the executed experiments of scenario 1 in the Workstation Cluster

Fast Ethernet Workstation Cluster	Number of processors =(Team Size x Number of Teams)			
	1 =(1x1)	4 =(2x2)	8 =(4x2)	16 =(4x4)
Average solution quality after 1000 seconds	4,487,343 100 %	4,211,036 93,84 %	4,161,165 92,73 %	4,089,542 91,14 %
Average number of moves per Team	449	1075	2212	2164

On account of the clustering of the entire network into small groups and the additionally applied dynamic load balancer, it was possible to efficiently use the available capacities in the workstation network. Note that besides the final exchange of the calculated results between the teams, communication operations are executed only within the teams throughout the computational process.

Owing to the combination of additional search paths and an accelerated move generation in each team, significant improvement of the attained solution quality can be achieved by the use of larger sized systems. Note that the solution quality was significantly worsened by the use of modified versions generating only a single path by grouping all workstations together within the same team. In addition to this, extremely diversified versions with teams comprising only a single computer and therefore generating p independent paths were significantly outperformed by the clustered version as well. Consequently, it can be concluded that this clustered version provides a more efficient combination for simultaneously using instruments of diversification and intensification. In addition, the simple load balancing scheme was able to perform an adaptive work sharing resulting in substantial improvement by the use of larger sized systems. On account of the comparatively poor sequential performance, it can be stated that sequential systems are overwhelmed with the complexity of the considered problem.

In order to evaluate the performance of the used dynamic load balancer in extreme background load scenarios, a second and more adversarial test environment was generated. Here, extreme background load changes occur instantly throughout the computations. This was initiated by the random launch of extremely time-consuming mathematical calculations. For this purpose, an independent background load thread commenced on each node in the network, started or broke up those calculations randomly. On account of the fact that these "applications" could even block entire computers, a simultaneous use by students was no longer possible. Consequently, these additional test scenarios were performed on a separate part of the network comprising in total only 8 processors. In order to evaluate the simple balancing scheme in more detail, a single path computation was examined there. Again, 10 different test instances were executed, whose average results after 1,000 seconds with and without the additional background load are depicted in the Tables 9-3 and 9-4.

Table 9-3. Results of a single searching path without artificial background load in scenario 2

	Number of Processors per Path		
	1	**4**	**8**
Average solution quality (after 1,000 seconds)	12,629,006	12,374,210	12,181,171
Average number of moves	143	574	1,061
Speedup	1	4.03	7.46

Obviously, without an existing artificial background load almost the full linear speedup was attained in average. Therefore, significant improvement of the average solution quality was attained by the analyzed searching path. In order to enable this, the heterogeneity of the background load caused by the working students was answered by moderate load balancing activities leading to a standard deviation of the overall total node work load caused by the Tabu Search procedure of 0.0807. Note that these results are average values which neglect extreme imbalances that may temporarily occur throughout the computation.

Table 9-4. The effect of the additional background load in the ETHERNET Cluster for a single searching path with 8 processors in scenario 2

Comparison: with background load / without background load	With additional background load	Without additional background load
Standard deviation of the Tabu Search workload	0.1321	0.0807
Standard deviation of the artificial mathematical background workload	0.3008	0
Standard deviation of the mean value of both workloads	0.1169	0.0807
Average number of executed moves	572.9	1061

By switching from scenario 1 (without artificial background load) to scenario 2 (with artificial background load), a significantly increased activity of the dynamic load balancer could be observed. This becomes directly visible through a comparison of the measured overall standard deviation of the Tabu Search workload. It increases now to 0.1321. On account of these activities taken by the load balancing scheme, the overall standard deviation of both (the artificial background load and the Tabu Search work) was 0.1169 which was 0.0514 larger than in the first scenario. But note that the difference of the deviation of the artificial mathematical background load was about 0.3008 and therefore considerably larger. In addition, in case of the scenario 1, the standard deviation of both workloads does not include the deviation of the background load caused by the applications started by the students.

Consequently, it can be stated that the simple dynamic load balancing scheme was able to balance the work distribution within the Tabu Search computation almost according to the current node performance. Note that owing to the fact that implementing a modified distribution needs no communication at all (i.e. no working package is exchanged between the affected nodes), this algorithm can cope with extremely adversarial scenarios where the work load extremely changes from one moment to another. On account of the fast move generation with an average duration of less than one second per iteration in a path of 8 computers, the algorithm can provide even under these circumstances a sufficiently fast load adaptation.

7.   CONCLUSIONS

This present paper provides a new approach for planning mixed-model assembly lines. By introducing a modular product architecture, which is common in variety steering and variety formation approaches, a direct interface to these program planning instruments is provided. Consequently, this offers the useful perspective for integrated and more comprehensive decision support. Thus, for deciding about integrating, retaining or eliminating specific offered product features within the variety steering decision processes, more detailed information according to the caused manufacturing costs can be respected. In addition, it has been shown that the resulting balancing approach can deal with theoretical variant programs comprising several billions of variants. Furthermore, available information according to the occurrence frequency of specific variant attributes can be used for finding a more appropriate layout constellation. Besides the problem definition for mixed-model balancing problems, the approach additionally comprises a specifically developed distributed Tabu Search procedure. In order to find appropriate layout constellations, it is executable in ordinary PC LANs which are available in almost all companies today. Owing to the existing unpredictable background load occurring in those companywide, heterogeneous networks, the Tabu Search algorithm used comprises a dynamic load balancer. In order to use the existing off-peak times of the connected computers, this balancer has to share the existing work of the search process according to the current performance of the network nodes. By analyzing the results measured for various test scenarios, it can be shown that despite its simplicity, the applied scheme was able to provide an adaptive balancing throughout the computations. On account of its structure which avoids any communication for the exchange of working packages, the algorithm retains its adaptability even in extreme adversarial scenarios. Altogether, it can be stated that despite an existing background load, the proposed scheme was able to attain considerable speedups and therefore significant improvement of the solution quality.

Future research motivated by these promising results should mainly proceed with the finding of a stronger connection between variety steering and line balancing. Specifically, a respective integrated hierarchical approach would require a detailed determination of a suitable interface between both decision processes. For instance, this interface has to define specific thresholds up to which a cost increase caused by an introduced product feature can be accepted. Obviously, for this purpose, the new approach provides an appropriate base. In particular, already during the variant program determination process, such an integrated approach would help to identify occurring bottlenecks and their consequences for the attainable efficiency. By avoiding substantial manufacturing problems and costs, this would be of particular importance. However, the generation of such a kind of interfaces is a very challenging field of research.

REFERENCES

Amen, M. 2005. Cost-oriented assembly line balancing: Model formulations, solution difficulty, upper and lower bounds. European Journal of Operational Research, in press.

Becker, C., Scholl, A. 2005. A survey on problems and methods in generalized assembly line balancing. European Journal of Operational Research, in press.

Blecker, T., Abdelkafi, N., Kaluza, B., Friedrich, G. 2003. Key Metrics System for Variety Steering in Mass Customization, in: Piller, F. T., Reichwald, R., Tseng, M. (eds.): Competitive Advantage Through Customer Interaction: Leading Mass Customization and Personalization from the Emerging State to a Mainstream Business Model. Proceedings of the 2nd Interdisciplinary World Congress on Mass Customization and Personalization - MCPC'03, pp. 1 - 27.

Blecker, T., Abdelkafi, N., Kaluza, B., Kreutler, G. 2004. A Framework for Understanding the Interdependencies between Mass Customization and Complexity. Proceedings of the 2nd International Conference on Business Economics, Management and Marketing, Athens / Greece, 1-15.

Blecker, T., Abdelkafi, N., Kaluza, B., Kreutler, G. 2004. Auction-based Variety Formation and Steering for Mass Customization. EM - Electronic Markets 14, 232 – 242.

Bock, S. 2000. Modelle und verteilte Algorithmen zur Planung getakteter Fließlinien. DUV, Wiesbaden.

Bock, S., Rosenberg, O. 2000. Dynamic Load Balancing Strategies for Planning Production Processes in heterogeneous Networks. Proceeding of the INFORMS Conference 2000 in Seoul, 951-955.

Bock, S., Rosenberg, O., van Brackel, T. 2005. Controlling mixed-model assembly lines in real-time by using distributed systems. European Journal of Operational Research, in press.

Bukchin, J. 1998. A comparative study of performance measures for throughput of mixed model assembly line balancing in JIT environment. International Journal of Production Research 36, 2669-2685.

Bukchin, J., Dar-El, E.M., Rubinovitz, J. 2002. Mixed model assembly line design in a make-to-order environment. Computers & Industrial Engineering 41, 405-421.

Bukchin, J., Masin, M. 2004. Multi-objective design of team oriented assembly systems. European Journal of Operational Research 156, 326-352.

Duray, R., Ward, P.T., Milligan, G.W., Berry, W.L. 2000. Approaches to mass customization: configurations and empirical validation. Journal of Operations Management 18, 605-625.

Fremery, F. 1991. Model-mix balancing: more flexibility to improve the general results. Production research: approaching the 21st century, Tailor & Francis, London, 314-321.

Hansen, P.; Mladenović, N. 2001. Variable neighborhood search: Principles and applications. European Journal of Operational Research 130, 449-467.

Macaskill, J.L.C. 1972. Production-line balances for mixed-model lines. Management Science 19, 423-434.

Piller, F., Koch, M., Moeslein, K., Schubert, P. 2003. Managing high variety: How to overcome the Mass Confusion Phenomenon. Proceedings of the EURAM 2003 Conference, Milan.

Pinnoi, A., Wilhelm, W.E. 1997. A family of hierarchical models for the design of deterministic assembly systems. International Journal of Production Research 35, 253-280.

Pinto, P.A., Dannenbring, D.G., Khumawala, B.M. 1983. Assembly Line Balancing with Processing Alternatives: An application. Management Science 29, 817-830.

Rosenberg, O. 1996. Variantenmanagement. In: Kern, W. et al. (Eds.). Handwörterbuch der Produktionswirtschaft, 2nd edition, Schäffer-Pöschel, Stuttgart, 2119-2129.

Rosenberg, O., Ziegler, H. 1992. A comparison of Heuristic Algorithms for Cost-Oriented Assembly Line Balancing. Methods and Models of Operations Research 36, 477-495.

Scholl, A. 1999. Balancing and Sequencing of Assembly Lines. 2nd edition, Physica, Heidelberg.

Scholl, A., Becker, C. 2005. State-of-the-art exact and heuristic solution procedures for simple assembly line balancing. European Journal of Operational Research, in press.

Thomopoulos, N.T. 1970. Mixed model line balancing with smoothed station assignments. Management Science 16, 593-603.

Chapter 10

ORDER FULFILLMENT MODELS FOR THE CATALOG MODE OF MASS CUSTOMIZATION – A REVIEW

Philip G Brabazon and Bart MacCarthy
Nottingham University Business School, Operations Management Division

Abstract: Catalogue Mass customizers are being imaginative in coping with the demands of high variety, high volume, customization and short lead times. These demands have encouraged the relationship between product, process and customer to be re-examined. A diversity of order fulfillment models are observed including some models with a single fixed decoupling point, some with multiple fixed decoupling points, and others with *floating* decoupling points.

Key words: Order fulfillment

1. INTRODUCTION

Mass Customization (MC) is not a mature business strategy. A range of operational approaches and order fulfillment models are being used in practice. From an examination of cases and empirical evidence, MacCarthy *et al* (2003) distinguished five fundamental operational modes for MC. The classification takes into account the way in which a firm's operational resources are used, whether or not the design envelope is predetermined and whether or not repeat orders are anticipated. Order fulfillment in any specific mode may be achieved in different ways. In some modes the order fulfillment process may vary depending on the requirements of a specific order.

One of the more common modes of Mass Customization - Catalogue MC - is the focus of attention in this paper. It is defined as the mode in which a customer order is fulfilled from a pre-engineered catalogue of potential variants that can be produced with a fixed order fulfillment process. In this mode product design and engineering are not linked to orders – they will have been completed before products are offered on the market and before orders are received. Likewise the order fulfillment activities will have been designed and engineered ahead of any order being taken.

In the Catalogue MC mode, customers select from a pre-specified product/variant/option range and the products are manufactured by the order fulfillment activities that are in place. This mode is relevant to consumer product markets and to many Business-to-Business environments for engineered products. If the design of the product is completed after consultation with the customer, or the order fulfillment system is modified for an order, then the mass customizing company is not operating in the Catalogue MC mode but in one of the other four modes.

Even when limiting the focus to the Catalogue MC mode there is no reason to believe that organizations are constrained to one model of how to achieve it operationally. Companies are approaching MC in general and the Catalogue mode in particular either from a mass production or a pure customization origin (Duray 2002). This in itself is reason to believe that several order fulfillment models will be observable in practice. In some cases it is variety proliferation that has motivated the uptake of Catalogue MC, for instance where variety is required for varied market segments, such as a global product that needs to be differentiated for different markets (Feitzinger & Lee 1997). In other cases it is customization for the end customer that is the motivation, such as computer servers (Swaminathan & Tayur 1998). The diversity of contexts is further reason to believe that a number of models are being and will need to be applied.

2.   DELINEATING THE ORDER FULFILLMENT PROCESSES

Order fulfillment is not a universally used term as noted by Kritchanchai & MacCarthy (1999) who found 'few sources in the literature discussing the details of the order fulfillment process explicitly'. There is no standard definition of order fulfillment and no common understanding of what activities it involves.

In the context of manufacturing, it is intuitive to say that order fulfillment involves the activities that lead to the hand-over of one or more ordered product to the customer. Beyond this it is less certain what activities should be treated as forming part of the order fulfillment process (OFP). To Shapiro *et al* (1992) the details vary from industry to industry but in general they see fulfillment as encompassing procurement, manufacturing, assembling, testing, shipping and installation. For them it does not include order entry. If the goal of order fulfillment is defined as complying with the customer's requirements i.e. the WHWW details (What product(s), How many, Where to deliver to, When to deliver) then the OFP is not involved directly with the customer to take the details of the order. Shapiro *et al* (1992) exclude order planning, order selection and scheduling from their fulfillment stage. Although they note that fulfillment can require considerable co-ordination, here we argue that to comply with the WHWW details not only must OFP encompass some material processing/transportation activities but also the key elements of control logic to plan and prioritize as well as co-ordinate activities.

At one extreme, where all the potential variety across the product range is pre-manufactured, the logic may be a simple rule for which product to take from stock and the activity be nothing more than handing it over to the customer. At the other extreme the OFP may involve the triggering and sequencing of complex production and distribution processes. While the details and scale of the OFP might differ greatly from one situation to another, in general terms the OFP encompasses the material processing activities concerned with complying with customer instructions and the control of these activities.

It is tempting to use the Customer Order Decoupling Point (CODP), which is 'traditionally defined as the point in the manufacturing value chain where the product is linked to a specific customer order' (Olhager 2003), to delineate the OFP. However, although activities upstream of the CODP will be controlled by forecasts, the state of these activities can have a bearing on the future performance of the downstream activities. This is particularly relevant when customer orders are conditional on delivery dates promised during sales negotiation. Although in some manufacturing systems the upstream and downstream activities may be insulated from each other, in general there are dependencies between them. If the customer's WHWW requirements are to be fulfilled, the OFP must have good situational awareness of the system – i.e. a grasp of the current state of the material processing activities, how they got into this state and, more importantly, how they are going to develop over time. For this to be the case the OFP cannot be blind to the upstream activities and, consequently, it is not appropriate to use the CODP as an OFP boundary marker.

The process of Demand Management, as described by Vollmann *et al* (1997) provides a template for defining and describing OFP. To Vollmann *et al* demand management is a highly integrative activity that captures and co-ordinates demand on manufacturing capacity. They note that 'the basic concept of demand management is that there is a pipe of capacity which is filled in the short run with customer orders and the long run with forecasts; order entry is a process of consuming the forecast with actual orders'. To them it encompasses forecasting, order entry, order-delivery-date promising, customer order service, physical distribution and other customer-contact-related activities.

2.1 Interpretation of order fulfillment and scope of the review

In this review, order fulfillment is interpreted in the following way:
- The OFP receives and acts upon customer orders, which contain the WHWW details (What product(s), How many, Where to deliver to, When to deliver);
- The OFP requires an awareness of the current and future state of the material processing activities. It envisages a pipeline of real and planned products and links customers to either type of product;
- The activities upstream of the CODP are within the bounds of the OFP if downstream activities are dependent on their performance.

Using this interpretation, literature that addresses the following issues with respect to Catalogue Mass Customization is considered to be relevant to the review:
- How products flows are structured in relation to processes, inventories and decoupling point(s);
- Characteristics of the OFP pipeline that inhibit or facilitate fulfillment;
- The logic of how products are allocated to customers;
- Customer factors that influence order fulfillment process design and operation.

3. ORDER FULFILLMENT STRUCTURES

The relative positions of processes and inventories are a fundamental aspect of order fulfillment models, as illustrated by Bucklin (1965) in his comparison of two different generic strategies - speculation and postponement. Compared to the speculation model, the stock of finished goods is not a feature of the postponement model

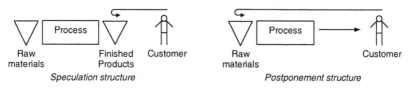

Figure 10-1. Speculation and Postponement structures (adapted from Bucklin, 1965)

The review of literature has identified four structural forms that have been claimed to be used in a Catalogue MC context (although authors do not use the term explicitly):

- Fulfillment from stock;
- Fulfillment from a single fixed decoupling point;
- Fulfillment from one of several fixed decoupling points;
- Fulfillment from several locations, with floating decoupling points.

3.1   Fulfillment from stock

Product variety has been on the increase in many sectors (Cox & Alm 1998). Since fulfillment of customers from stock is still prevalent it is unsurprising that examples can be found that claim to have adapted this configuration to high variety / mass customization situations.

It can be debated whether or not stock fulfillment models should be included in the review and which papers should qualify for inclusion. For example, the customization of printers by Hewlett Packard (Feitzinger & Lee 1997, Lee & Billington 1995) is heralded as mass customization but the end consumer is not involved in the process. Customization is required for the region in which the printer is to be sold and hence the study could be relabeled as solely a case-study in postponement. There is an argument that it illustrates postponed manufacture and that the label of mass customization was given to it before MC strategies were more widely scrutinized. It is included here, with two other examples, to show the diversity of approaches for coping with high product variety. The three examples are summarized below and the figure overleaf:

- Hewlett-Packard printers are customized for each region by postponing some assembly and packaging activities. Standard unfinished units are shipped from a central facility to each region for completion (Feitzinger & Lee 1997, Lee & Billington 1995).
- In the context of the automotive sector Boyer & Leong (1996) study a structure in which multiple product types are supplied to many stock locations. They study the impact on the system of increasing the number of products that each plant can produce.
- Herer *et al* (2002) examines the method of transshipment for a high variety of products, which is the ability to transfer stock between locations at the same echelon level. Transshipment is a form of physical postponement and as Herer puts it, creates the ability to transform a generic item (an item at any location) into a specific item (an item at a specific location) in a relatively short time.

A theme of the research into stock fulfillment structures is how to structure the processes that replenish the stock to cope with variety without suffering high costs. Hau Lee is one of the principal contributors in this area and he sees a key issue to be how product design interacts with the process (e.g. Lee & Billington 1995, Lee 1996, Lee & Tang 1997, Lee & Tang 1998, Whang & Lee 1998). Whang & Lee (1998) present models to indicate the scale of benefit that postponement can bring through uncertainty reduction and reduced forecasting error. Lee & Tang (1997) use a model to study three approaches to delaying product differentiation, taking forward the models of Lee (1996). Lee & Tang (1998) study further the approach of operations reversal and put forward properties that an order fulfillment sequence should strive for when the major source of demand uncertainty lies in the option mix and the total demand for all options is fairly stable.

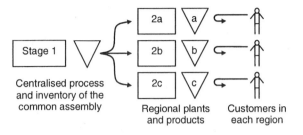

Adapted from Feitzinger & Lee (1997)

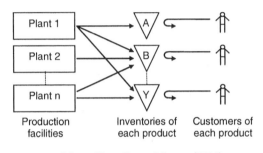

Adapted from Boyer & Leong (1996)

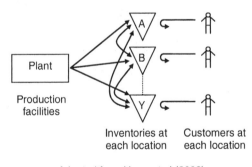

Adapted from Herer et al (2002)

Figure 10-2. Structures for fulfillment from stock

3.2   Fulfillment from a single fixed decoupling point

This structure takes the form of the postponement model described by Bucklin (1965) and others subsequently (van Hoek 2001, Yang and Burns 2003). Of the four types of OFS structures this is the format that tends to be associated with catalogue mass customization. In this structure the producer holds stocks of raw materials or part-finished products and once an order is received these are taken forward to be completed and delivered to the customer.

A commonly used standard classification of order fulfillment systems includes a set of fixed decoupling point structures: engineer-to-order (ETO); make-to-order (MTO); and assemble-to-order (ATO). Hill (1995) extended this by adding design-to-order and make-to-print. Recently, the category of configure-to-order (CTO) has been distinguished as a special case of assemble-to-order (Song & Zipkin 2003), in which components are partitioned into subsets from which customers make selections (e.g. a computer is configured by selecting a processor from several options, a monitor from several options, etc).

Many practicing mass customizers have a single fixed decoupling point and fit into the assemble-to-order or configure-to-order categories, though they may also perform some fabrication activities in the customization they offer:

- Kotha (1995) describes the Japanese bicycle company, National Panasonic, who await each order before fabricating the frame and assembling the bicycle with components from stock;
- A series of articles describe how the UK company RM switched its computer supply business from a make-to-stock to an assemble-to-order fulfillment mode (Duffel 1999, Duffel & Street 1999).
- Orangebox is a UK company producing office furniture. Their products are modular and they produce high levels of variety in small batch sizes. Once an order is received they cut and sew the covers and assemble the product from components in stock (Tozer 2003).

In a high product variety environment, Dobson & Stavrulaki (2003) analyze the switch to a finish-to-order strategy from MTS strategy. The company produced dozens of intermediate products that were then cut into hundreds of sizes and shapes. Orders were received from many dealers dispersed across the US. Although set-up costs rose, the switch was beneficial in terms of stock costs.

3.3 Fulfillment from one of several fixed decoupling points

These structures have more than one decoupling point, i.e. there are two or more distinct stock holding locations within the production and delivery processes from which raw materials or part-finished products can be taken, allocated to a customer, finished and delivered. A customer need not be aware of which decoupling point is being used for their order.

Graman & Magazine (2002) study an OFP with two fixed decoupling points – one is mid process and the other is the finished stock. They conclude that holding some items in a part-finished state and retaining some final processing capacity open to fulfill orders can bring significant performance benefits, compared to a situation in which all orders are filled from stock.

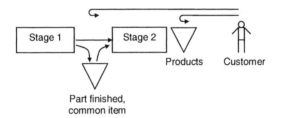

Figure 10-3. Structure studied by Graman & Magazine (2002)

For an Integrated Steel Mill (ISM), Denton *et al* (2003) develop a model to select which steel slabs to hold in a mid-process inventory. Competitive pressure from mini-mills has been forcing ISMs to shift to the high-end markets for exotic or custom-finished steel products. ISMs used to operate in the MTO production mode, with processes designed for high volumes and order fulfillment times in the range of 10 to 15 weeks, but the high-end markets demand custom products with shorter and reliable delivery lead times, in the range of 5 to 6 weeks. The result of increased product variety has been capacity shortages and exploding inventories. Prior to implementing their model, 57 slab designs covered about 17 percent of total annual order volume, and after implementation 50 slab designs covered about 50 percent of the total annual order volume.

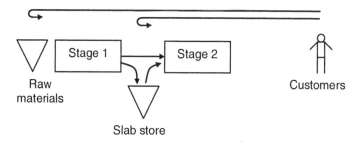

Figure 10-4. Structure studied by Denton *et al* (2003)

Swaminathan & Tayur (1998) study an OFP with three fixed decoupling points – one at the start of final assembly, one mid-assembly and also finished stock. They develop a model to tackle a problem in which a producer offers a broad product range but in each time period orders are received for a fraction of variants only. They compare a *vanilla box* strategy (in which subsets of components are pre-assembled into a number of vanilla boxes, that can be used for a number of variants, exploiting the inherent commonality in the product family) against MTS and ATO strategies (and mixes of the three) and find the vanilla box approach can be superior significantly. In exploring their model, they show how factors including capacity constraints, demand correlation, number of vanilla box types and breadth of product range alter the performance of each strategy. In a second study, Swaminathan & Tayur (1999) go on to develop models that take account of assembly precedence constraints, in particular the feasibility of a vanilla box in terms of whether it can be assembled.

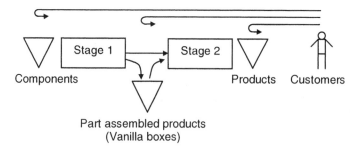

Figure 10-5. Structure studied by Swaminathan & Tayur (1998, 1999)

3.4   Fulfillment from several process points, with floating decoupling points

The key feature of order fulfillment systems with this structure is that products can be allocated to orders at any point along the process, hence the coining here of the term *floating decoupling point*. This structure is observed in the capital goods sector but is being adopted elsewhere including the automotive sector.

Manufacturers of complex goods with relatively long production lead times, such as machine tools, have been facing the challenges of increased product diversity and shortening of delivery lead times. The requested delivery lead time is often less than the sum of purchasing, fabrication and assembly lead times. As a consequence such companies have been evolving their order fulfillment processes. In their study of three heavy manufacturing firms Raturi *et al* (1990) describe how firms have implemented a build-to-forecast (BTF) schedule in which they forecast end-product mix, create a master schedule of end-products and then release production orders before specific customer orders are received.

In BTF there is no stopping point in the production process so mid process buffer inventories are avoided. Customer orders are matched to items in any state of production that will meet the due date. Customer orders rarely match the end products being built hence orders are fulfilled by:

- changing products early in the process if the basic model is an appropriate one and the production plan can be altered to accommodate the actual order;
- reconfiguring an end product, with features removed and replaced as required. On occasion the changes are so extensive that a loss is incurred.

Bartezzaghi & Verganti (1995a, b) study a market for capital telecommunications equipment in which there are few but large and powerful buyers. A manufacturer expects contracts to be issued but the volume and specification is uncertain, and some degree of manufacturing must be initiated so as to be in a position to meet the delivery schedule.

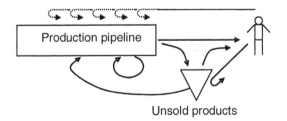

Figure 10-6. Structure studied by Raturi *et al.* (1990)

The development of information systems has led automotive fulfillment processes to evolve into a floating decoupling point structure. The multi-mechanism system has been labeled *Virtual Build-to-Order* (VBTO) and Agrawal *et al* (2001) describe it as connecting customers:

> 'either via the internet or in dealer's showrooms, to the vast, albeit far-flung, array of cars already in existence, including vehicles on dealer's lots, in transit, on assembly line, and scheduled for production',

with the expectation that:

> 'customers are likely to find a vehicle with the color and options they most want'.

Holweg (2000) also describes in the automotive context the multiple fulfillment mechanisms by which a customer can receive a vehicle: from the local dealer's stock; by a transshipment from another dealer's stock; by a vehicle taken from a central stock holding centre; by a vehicle being submitted into the order bank as a build-to-order product; or by a vehicle that is in, or scheduled for, production being allocated to the customer, which may involve its specification being amended.

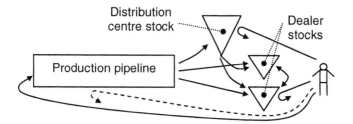

Figure 10-7. Structure studied by Holweg (2000)

A generic VBTO model of this type has been studied in detail by Brabazon and MacCarthy (2004a, b, 2005). This work has identified reconfiguration flexibility as being a key capability of the system. The greater the flexibility the greater the likelihood a customer can be matched to a product in the production and distribution pipeline. Their simulation based studies have shown an unexpected result in that the VBTO fulfillment system has a propensity to cause average stock levels to rise even when production and demand are harmonized.

4.   INFLUENCES ON THE CHOICE OF FULFILLMENT MECHANISM

4.1   Product variety

Product variety is increasing, evidenced by data (Cox & Alm 1998) and by the initiation of research into the challenges created by product variety (Ramdas 2003). High levels of product variety tend to go hand in hand with a mass customization strategy (e.g. Denton *et al* 2003, Swaminathan and Tayur 1998).

From the literature it is evident that product variety creates challenges for the design and operations functions (see the review by Ramdas 2003). Even when not approaching the issue from a customization context, much of the literature on product variety has some relevance to order fulfillment. However, the aim here is not to review a swathe of literature and draw out general lessons about variety as such reviews have been done. The aim is to review literature that has addressed product variety in an MC context. A problem with this ambition is that most, if not all of the papers reviewed in this chapter could be classified as addressing product variety. In the end, two papers are identified that deal with variety and that qualify for inclusion.

McCutheon *et al* (1994) discuss the implications of what they term as the 'customization-responsiveness squeeze' to manufacturers of capital equipment, and machine tools in particular. Among the coping tactics for a manufacturer, they note: increasing the flexibility of the process; altering the product design (i.e. designing for postponement); managing demand including how sales link to the production planning process; and following a build-to-forecast (BTF) approach.

Salvador & Forza (2004) also look at implications of the customization-responsiveness squeeze to the whole enterprise and find that customized variety creates difficulties in the areas of sales, technical office functions (e.g. documentation) and manufacturing.

4.2   Postponement

Delaying the completion of the product until a customer order is received is a tactic used in many MC applications summarized above.

A recent review of postponement distinguishes several forms: product development postponement, purchasing postponement, production postponement and logistics postponement (Yang & Burns 2004). All forms are relevant to MC but product development postponement is not relevant to

Catalogue MC as this is a mode in which the product is fully engineered before the customer order is taken.

Postponement and a modular product are two approaches that are presented often as being essential for MC. For example, Partanen & Haapasalo (2004) state:

'The fundamental idea behind mass customization and modularization is that the order penetration point is delayed as late as possible.'

However, the review of fulfillment models above identified MC applications where customers are fulfilled from stock as well as from part-finished products. These applications refute the claim that postponement is essential.

4.3   Process flexibility

In environments of high product variety and customization, the characteristic of *flexibility* is highlighted in the literature as being the key facilitator/inhibitor. Several sources consider flexibility to be an enabler of mass customization (Fogliatto *et al* 2003, Da Silveira *et al* 2001, Kakati 2002, Duffell & Street 1998) and the ability to be flexible is assumed within analyses of the economic impact of mass customization (de Vaal 2000, Norman 2002). A wider range of products and increased customization are identified by De Toni & Tonchia (1998) as two of five motivations for flexibility, the others being: variability of demand (random or seasonal); shorter life-cycles of the products and technologies; and shorter delivery times.

There is a considerable body of flexibility research but the breadth of the topic is vast with the topic being approached at one end as a concept, and at the other it is examined in the context of a specific situation. The scale of concern ranges from the flexibility of a sector down to the flexibility of a machine or fixtures and the concept also has a temporal property – flexibility over a short or long time horizon. Although there is a large volume of literature on flexibility, there is little that focuses on the flexibility of mass customization systems specifically. This is not to say that flexibility has not been of interest in mass customization research, but it has not been the focus.

Several studies have been identified that assess flexibility and are relevant to catalogue mass customization.

Bradley & Blossom (2001) estimate the change in cost and the improvement in delivery lead time that would be achieved by an assembly process if it were to accept a higher proportion of customer orders. The study is in the automotive sector and the order fulfillment process under consideration resembles a floating decoupling point system. The study does not look at how customer orders are matched to vehicles in the pipeline, but recognizes this is an area that needs attention. Their supposition is that flexibility can be increased in the assembly line by adding production capacity (people or equipment) so that a fluctuating mix of products can be produced. Thus the products made on the assembly line can be those that the customers want, when they want them, rather than units selected for attainment of maximum efficiency. By simulating a generic automotive system they estimate, in the worst case, cost would rise by around 0.017% at a level of 70% make-to-order (a significant reason being that direct labor accounts for only 6% of costs typically) and delivery lead times would reduced by around 70%.

Bukchin *et al* (2002) develop a heuristic for designing assembly lines for mixed model operations. They assume the model-mix is determined ahead of time and stable (say for a year ahead) but the sequence of launching products to the line must be determined by actual short range demand patterns and customer orders. Their approach assumes a model mix for which the combined workload is balanced for the duration of the entire shift and not on the basis of station cycle times (as was the case for single model assembly).

Boyer & Leong (1996) develop a model for evaluating the benefit of increasing levels of flexibility. Their context is the automotive sector and the point of interest is the ability for a number of plants to produce more than one product line. Without flexibility, unused capacity in one plant cannot be used to fulfill demand that exceeds the capacity of another. They find that opening up a fraction of the feasible cross-links between products and plants brings substantial gains in overall performance, even with a throughput loss of 20% due to changeovers.

To counter supply chain effects, the Quantity Flexibility (QF) contract has become popular (Tsay & Lovejoy 1999). It attaches a degree of commitment to the forecasts by installing constraints on the buyer's ability to revise them over time. The extent of revision flexibility is defined in percentages that vary as a function of the number of periods away from delivery. The QF contract formalizes the reality that a single lead time alone is an inadequate representation of many supply relationships, as evinced by the ability of buyers to negotiate quantity changes even within quoted lead times. The model indicates that inventory is a consequence of disparities in flexibility. In particular, inventory is the cost incurred in overcoming the inflexibility of a supplier to meet a customer's desire for flexible response and they coin the term *flexibility amplification*. All else being equal, increasing a supply chain participant's input flexibility reduces its costs, promising more output flexibility comes at the expense of greater inventory costs.

4.4   Fulfillment logic

The issue of how to link orders to products or production capacity is a key aspect of Demand Management. Rules such as assigning orders to the 'earliest available' and 'latest available' production slot have been examined (e.g. Guerrero 1991). The *production seat system* (Tamura & Fujita 1995, Tamura *et al* 1997, Tsubone & Kobayashi 2002) is a demand management system for producers of a variety of complex products in mixed or small lots, developed for the purpose of shortening delivery lead times. It deliberately creates capacity slots of different dimensions in recognition of differences in product manufacturing requirements and the sales team can see which slots are free when negotiating with the customer.

In the context of floating decoupling point systems, Brabazon & Mac-Carthy (2004b) study how search rules alter the likelihood of finding a match for a customer. Their models identify the ratio of product variety to pipeline length as being an indicator of fulfillment performance of a system.

In their study using the vanilla box concept, Swaminathan and Tayur (1998) make the conjecture that it could be cost effective for the producer to supply the customer with a product that has superior grades of component(s) or even includes redundant components that the customer may not be made aware of, if the consequence of not doing so is to lose the sale. Giving customers substitutions when there are component or capacity shortages is not a new idea but, as Ramdas (2003) comments.

'there has overall been little research that addresses component-sharing issues for components with a strong influence on consumer perceptions.'

Balakrishnan & Geunes (2000) is one of the many studies that examines substitution without distinguishing between components that have strong and weak influence on consumer perceptions, but a useful concept they introduce in their analysis is *conversion costs*:

'the per-unit conversion cost represents any additional processing effort or cost incurred when we substitute a preferred component with an alternate component.'

The concept is not taken further in their analysis, but it is a concept of interest and is consistent with the concept of reconfiguration cost analyzed by Brabazon & MacCarthy (2004a).

4.5   Customer factors

Differences across customers are, of course, the prime reason for the growth in product variety. However, customer differences can be expected to create other forms of 'service' variety within the order fulfillment system, such as variety in lead time and price.

Price and lead-time are interrelated. Price is connected to value (e.g. Meredith *et al* 1994) and it is well understood that value tends to decay over time (e.g. Lindsay & Feigenbaum 1984). However, the rate of decay is not uniform across customer groups and for some customers, delivery earlier than an agreed date is undesirable.

Methodologies for exploiting customer differences are now emerging under the banner of yield management (also known as revenue management) and its proponents see many opportunities for exploiting its principles (Marmorstein *et al* 2003) but the research in the x-to-order sector is scarce. Tang & Tang (2002) study time-based policies on pricing and lead-time for a build-to-order and direct sales manufacturer of products for which value decreases rapidly with time, such as is the case with high technology components. Although Chen (2001) is not focusing on high variety systems, his work is relevant. He proposes customers be given the opportunity to select from a menu of price and lead time combinations, with greater price discounts on offer for longer delivery times. By reducing the proportion of customers who demand immediate fulfillment, his model shows how safety stock/inventory along a supply system can be reduced.

Evidence from the automotive domain supports the view that customers vary in their attitude and willingness to waiting. From a survey of customers who had bought new cars, Elias (2002) plotted distributions of the importance of waiting time in the purchase decision and of the customer's ideal waiting time.

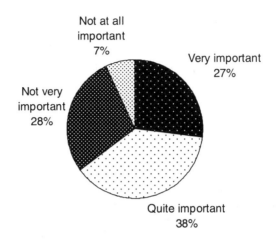

Figure 10-8. Importance of wait in the decision to buy, Elias (2002)

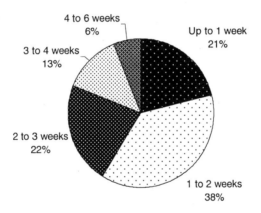

Figure 10-9. Ideal wait from placing the order to delivery, Elias (2002)

Customers can be expected to differ in their attitudes toward specification compromise as well as to delivery time. In an ATO context Iravani *et al* (2003) use simulation to find that customer tolerances to substitutions have an impact on the stock policy of an ATO system. They divide customers into four groups that differ in regard to the components that are key and non-key, and the substitutions they are prepared to accept (e.g. a proportion of customers, m, is prepared to accept item B instead of A with 1-m lost sales, and if B is also unavailable a proportion, n, will take C and m-n will also be lost). In their system, customers must get their key items or acceptable substitutions for their key items, but will still make a purchase if a non-key item is unavailable and cannot be substituted. They use several overlapping measures:

- Fully satisfied – customers getting the exact match for key and non-key components;
- Key satisfied – customers who get all of their key items, but some or all of their non-key items are substituted (note Fully satisfied is a subset of this category);
- Substitution satisfied – customers who accept a substitution for at least one of their key items.

A similar approach to segmenting customers is used by Brabazon & MacCarthy (2005). They assume every customer is seeking a target specification but that customers will treat each feature as being *critical* or *non-critical:* a critical feature is one for which the customer must receive their target option; but the customer will tolerate an alternative option for a non-critical feature. The proportion of each customer type is varied, revealing the sensitivity of fulfillment metrics to the mix.

5. CONCLUSIONS

Diversity in the order fulfillment structures was expected and is reported in the literature. Product variety has been increasing (Cox & Alm, 1998). Mass customization is not a mature operations model and hence diversity in operations models can be expected. What is apparent is that producers are being imaginative in coping with the demands of high variety, customization and short lead times. Such demands require and have encouraged the relationship between product, process and customer to be re-examined. Not only has this strengthened interest in commonality and postponement, but, as shown here, has led to the re-engineering of the order fulfillment process to create models with multiple fixed decoupling points and the floating decoupling point system.

Catalogue MC is one of the fundamental operational modes for Mass Customizing enterprises and is one of the most common types of MC occurring in practice. The review demonstrates that it may be achieved in a variety of ways with different order fulfillment configurations and operating policies. The work of the authors is evaluating the factors that influence the choice of OFP in Catalogue MC mode.

There are many avenues worthy of research. Market conditions and technology are driving the re-engineering of the order fulfillment process but there remains the question as to how these structures and their control logic perform and under what circumstances they offer benefits, in particular where there are customer differences, not just in terms of time and cost trade-offs that are exploited in revenue management but in terms of product features.

REFERENCES

Agrawal, M., Kumaresh, T. & Mercer, G. (2001). The false promise of mass customization. The McKinsey Quarterly, 3, 62-71Agrawal, M., Kumaresh, T. & Mercer, G. (2001). The false promise of mass customization. The McKinsey Quarterly, 3, 62-71.

Balakrishnan, A. & Geunes, J. (2000). Requirements Planning with Substitutions: Exploiting Bill-of-Materials Flexibility in Production Planning, Manufacturing & Service Operations Management, 2 (2) 166-185.

Bartezzaghi, E. & Verganti, R. (1995). Managing demand uncertainty through order over-planning. International Journal of Production Economics, 40, 107-120.

Bartezzaghi, E. & Verganti, R. (1995). A technique for uncertainty reduction based on order commonality. Production Planning & Control, 6, 157-170.

Boyer, K. K. & Leong, G. K. (1996). Manufacturing flexibility at plant level. International Journal of Management Science, 24, 495-510.

Brabazon, P. G. & MacCarthy, B. L. (2004a). Virtual-build-to-order as a Mass Customization order fulfilment model. Concurrent Engineering: Research and Applications, 12, 155-165.

Brabazon, P.G. & MacCarthy, B.L. (2004b). Fundamental behaviour of Virtual-Build-to-Order systems. Thirteenth International Working Seminar on Production Economics, February 16-20, Igls, Innsbruck.

Brabazon, P.G., MacCarthy, B.L. & Hawkins, R.W. (2005). Implications of customer differences and reconfiguration flexibility on fulfilment performance in a virtual-build-to-order system. 18th International Conference on Production Research, July 31 – August 4, Salerno, Italy.

Bradley, J.R. & Blossom, A.P. (2001) Using Product-Mix Flexibility to Implement a Make-to-Order Assembly Line, In: INFORMS International, Hawaii.

Bucklin, L. (1965). Postponement, speculation and the structure of distribution channels. Journal of Marketing Research, 2, 26-31.

Bukchin, J., Dar-El, E. M. & Rubinovitz, J. (2002). Mixed model assembly line design in a make-to-order environment. Computers & Industrial Engineering, 41, 405-421.

Chen, F. (2001). Market Segmentation, Advanced Demand Information, and Supply Chain Performance. Manufacturing & Service Operations Management, 3, 53-67.

Cox, W. M. & Alm, R. (1998). The right stuff: America's move to mass customization. Annual Report, Federal Reserve Bank of Dallas.

Da Silveira, G., Borenstein, D. & Fogliatto, F. S. (2001). Mass customization: Literature review and research directions. International Journal of Production Economics, 72, 1-13.

De Toni, A. & Tonchia, S. (1998). Manufacturing flexibility: a literature review. International Journal of Production Research, 36, 1587-1617.

Denton, B., Gupta, D. & Jawahir, K. (2003). Managing Increasing Product Variety at Integrated Steel Mills, Interfaces, 33, 2 41-53.

Dobson, G. & Stavrulaki, E. (2003). Capacitated, finish-to-order production planning with customer ordering day assignments, IIE Transactions, 35, 5, 445-455.

Duffell, J. (1999). Mass customisation across the business: Part 1. Control, 24, 9-11.

Duffell, J. & Street, S. (1999). Mass customisation across the business: Customized production Part 3. Control, 25, 24-26.

Duray, R. (2002). Mass customization origins: mass or custom manufacturing? International Journal of Operations & Production Management, 22, 314-328.

Economist (2001). Wave goodbye to the family car. The Economist, 358, 8204, 57-58.

Elias, S. (2002) 3 day car programme: New car buyer behaviour, Cardiff University.

Feitzinger, E. & Lee, H. L. (1997). Mass customization at Hewlett-Packard: the power of postponement. Harvard Business Review, 75, 116-121.

Fogliatto, F. S., Da Silveira, G. & Royer, R. (2003). Flexibility-driven index for measuring mass customization feasibility on industrialized products. International Journal of Production Research, 41, 1811-1829.

Graman, G. A. & Magazine, M. J. (2002). A numerical analysis of capacitated postponement. Production and Operations Management, 11, 340-357.

Guerrero, H. H. (1991). Demand management strategies for assemble-to-order production environments. International Journal of Production Research, 29, 39-51.

Herer, Y. T., Tzur, M. & Yucesan, E. (2002). Transshipments: An emerging inventory recourse to achieve supply chain leagility. International Journal of Production Economics, 80, 201-212.

Hill, T.J. (1995). Manufacturing Strategy: Text and cases. Basingstoke: Macmillan.

Holweg, M. (2000). The order fulfilment process in the automotive industry. Lean Enterprise Research Centre, Cardiff Business School.

Iravani, S. M. R., Luangkesorn, K. L. & Simchi-Levi, D. (2003). On assemble-to-order systems with flexible customers. IIE Transactions, 35, 389-403.

Kakati, M. (2002). Mass customization - needs to go beyond technology. Human Systems Management, 21, 85-93.

Kotha, S. (1995). Mass Customization: Implementing the emerging paradigm for competitive advantage. Strategic Management Journal, 16, 21-42.

Kritchanchai, D. & MacCarthy, B. L. (1999). Responsiveness of the order fulfilment process. International Journal of Operations and Production Management, 19, 812-834.

Lee, H. L. (1996). Effective inventory and service management through product and process redesign. Operations Research, 44, 151-159.

Lee, H. L. & Billington, C. (1995). The Evolution of Supply-Chain-Management Models and Practice at Hewlett-Packard. Interfaces, 25, 42-63.

Lee, H. L. & Tang, C. S. (1997). Modelling the costs and benefits of delayed product differentiation. Management Science, 43, 40-53.

Lee, H. L. & Tang, C. S. (1998). Variability reduction through operations reversal. Management Science, 44, 162-172.

Lindsay, C. M. & Feigenbaum, B. (1984). Rationing by Waiting Lists. American Economic Review, 74, 404-417.

MacCarthy, B. L., Brabazon, P. G. & Bramham, J. (2003). Fundamental modes of operation for mass customization. International Journal of Production Economics, 85, 289-304.

Marmorstein, H., Rossomme, J. & Sarel, D. (2003). Unleashing the power of yield management in the internet era: Opportunities and challenges. California Management Review, 45, 147-167.

Meredith, J. R., McCutcheon, D. M. & Hartley, J. (1994). Enhancing Competitiveness through the new market value equation. International Journal of Operations and Production Management, 14, 7-22.

Norman, G. (2002). The relative advantages of flexible versus designated manufacturing technologies. Regional Science and Urban Economics, 32, 419-445.

Olhager, J. (2003). Strategic positioning of the order penetration point. International Journal of Production Economics, 85, 319-329.

Partanen, J. & Haapasalo, H. (2004). Fast production for order fulfillment: Implementing mass customization in electronics industry, International Journal of Production Economics, 90, 2, 213-222.

Ramdas, K. (2003). Managing product variety: an integrative review and research directions, Production & Operations Management, 12, 1, 79-101.

Raturi, A. S., Meredith, J. R., McCutcheon, D. M. & Camm, J. D. (1990). Coping with the build-to-forecast environment. Journal of Operations Management, 9, 230-249.

Salvador, F. & Forza, C. (2004) Configuring products to address the customization-responsiveness squeeze: A survey of management issues and opportunities, International Journal of Production Economics, 91, 3, 273-291.

Shapiro, B.P., Rangan, V.K. & Sviokla, J.J. (1992). Staple Yourself to an Order, Harvard Business Review, 70, 4, 113-122.

Song, J.-S. & Zipkin, P. (2003). Supply Chain Operations: Assemble-to-Order and Configure-to-Order Systems. In S. C. Graves & A. G. e Kok (Eds.) Handbooks in Operations Research and Management Science: Design and Analysis of Supply Chains Elsevier Science Publishers (North Holland).

Swaminathan, J. M. & Tayur, S. R. (1998). Managing Broader Product Lines through Delayed Differentiation Using Vanilla Boxes. Management Science, 44, S161-S172.

Swaminathan, J. M. & Tayur, S. R. (1999). Managing design of assembly sequences for product lines that delay product differentiation. IIE Transactions, 31, 1015-1026.

Tamura, T. & Fujita, S. (1995). Designing customer oriented production planning system (COPPS). International Journal of Production Economics, 41, 377-385.

Tamura, T., Fujita, S. & Kuga, T. (1997). The concept and practice of the production seat system. Managerial and Decision Economics, 18, 101-112.

Tang, K. & Tang, J. (2002). Time-based pricing and leadtime policies for a build-to-order manufacturer. Production and Operations Management, 11, 374-392.

Tozer, E. (2003). Mass Customizing office furniture at Orangebox. Mass Customization: Turning customer differences into business advantage, IEE seminar 2003/10031, London.

Tsay, A. A. & Lovejoy, W. S. (1999). Quantity Flexibility Contracts and Supply Chain Performance. Manufacturing & Service Operations Management, 1, 89-111.

Tsubone, H. & Kobayashi, Y. (2002). Production seat booking system for the combination of make-to-order and make-to-stock products. Production Planning & Control, 13, 394-400.

van Hoek, R.I. (2001) The rediscovery of postponement: a literature review and directions for research, Journal of operations management, 19, 2, 161-184.

Vollman, T.E., Berry, W. & Whybark, D.C. (1988). Manufacturing planning and control. Homewood, IL: Dow Jones Irwin.

Whang, S. & Lee, H. L. (1998). Value of Postponement. In T.-H. Ho & C. S. Tang (Eds.) Product variety management: research advances (pp. 65-84). Boston: Kluwer Academic.

Yang, B. & Burns, N. (2003) Implications of postponement for the supply chain, International Journal of Production Research, 41, 9, 2075-2090.

Yang, B., Burns, N.D. & Backhouse, C.J. (2004). Management of uncertainty through postponement, International Journal of Production Research, 42, 6, 1049-1064.

Chapter 11

CUSTOMER SERVICE LEVEL IN A LEAN INVENTORY UNDER MASS CUSTOMIZATION

Wuyi Lu, Janet Efstathiou and Ernesto del Valle Lehne
University of Oxford, Department of Engineering Science

Abstract: Mass customization (MC) aims to satisfy customers' diverse demands with attractive product prices and short customer waiting time by adopting advanced management and manufacturing technologies. Theoretically, mass customization does not permit any stock of finished goods (FG). This optimal target can be achieved if there is a rapid and efficient supply chain to respond to the demand and production instantaneously. However, uncertainty and unpredictability from customers, suppliers and manufacturing systems make it a challenging goal. If there is no FG inventory available for some impatient customers, lost sales may result and customer service level will be affected.

Customers become involved in an MC facility at the decoupling point, where the push line gives way to the customizing, pull line. At this point, maintaining a small stock of semi-finished goods of the most popular product variants is beneficial for quickly customizing and delivering to the customers. However, we need to pay careful attention to these semi-finished and finished goods inventories because of the high holding and maintenance costs, and the obsolescence and depreciation of stocks.

In Lean manufacturing, inventory is minimized to avoid waste. Following this principle, we design a lean inventory for mass customization that does not consume too many resources or physical spaces. This inventory should have a product variety sufficient to meet customers' rapid evolving preferences.

We model this lean inventory consisting of n inventory locations, with m different product variants stored in the inventory. We assume an inventory location has equal probability $1/m$ to hold any one of these m variants. We define the lean inventory as dynamic because a customer arrival pattern and replenishment policies are introduced into the inventory. We assume there are one customer arrival and one delivery in each time step or each cycle. Within one cycle, a customer arrives and searches for their desired variant. If the variant is found, it is withdrawn and is replaced by another randomly selected variant. If the variant is not found, the sale is lost and there is no replacement.

In this paper, our research objective is to investigate the customer service level of lean inventory under mass customization. Customer service level is defined as the average probability of finding a customer's desired variant from the lean and dynamic inventory. We successfully apply combinatorial mathematics in developing a mathematical expression for calculating this probability. This will enable inventory practitioners to understand the relationship between customer service level, number of inventory locations (inventory capacity) and number of product variants (product variety). For example, if inventory managers want to achieve targeted customer service level, they can balance inventory capacity and product variety using our expression. They can either increase inventory capacity and have more inventory units, or decrease product variety and have fewer kinds of variants in the inventory.

Key words: Lean Inventory; Customer Service Level; Combinatorial Mathematics.

1.   INTRODUCTION AND OUTLINE

Mass customization (MC) challenges current manufacturing strategy to adopt advanced management and manufacturing technologies to cut down cost from production and inventory, and to shorten customer waiting time, without losing sales [1]. Pure mass customization does not permit any stock of finished goods [2]. However, there are two possible locations where some inventory might be required in an MC facility: at the decoupling point [3], or customer involvement point, where the push line gives way to the customizing, pull line, and as a small stock of finished goods of the most popular variants. The design of the mass-customizing factory needs to pay careful attention to the location of any inventory, not only because of the costs involved but also the rapid evolution of customer preferences.

In designing an MC facility, the inventory must be located, whether at the decoupling point or as finished goods, in order to be able to retrieve items and customize them to the customers' requirements at a lead-time and cost that are acceptable to the customers. The decoupling point and the product structure must be designed carefully. Many advocates of mass customization recommend placing the decoupling point as late as possible in the manufacturing process, so as to shorten the time required from the customer placing the order to delivery of the customized product. The push line can make semi-finished variants, which may be held in stock for rapid retrieval, customization and delivery to the customer. However, holding a large stock of many kinds of semi-finished variants runs into the problems of holding costs, obsolescence and depreciation. Holding too few means that the customer's order cannot be met rapidly from the inventory of semi-finished goods. Lean manufacturing advocates the elimination of waste and delay, so we follow those aims by designing an inventory strategy that does not consume too many resources and keeps the holding and maintenance costs as low as possible. Here, we consider a lean inventory with limited capacity and a fixed number of variants held in stock. We can observe such inventory in car dealerships and supermarkets with a limited stock of diverse products.

The problem addressed in this paper is to investigate the probability of meeting the customer's requirement when holding n items of stock, consisting of m different finished or semi-finished variants. The inventory is dynamic because customers withdraw their selected variant from stock randomly, and the inventory is replenished within the same cycle with a random variant.

The outline of the paper is as follows. In Section 2, we develop a model of a lean, dynamic inventory. Section 3 presents a mathematical model for this inventory. Section 4 discusses the relationship between customer service level, inventory capacity and product variety. We summarize and give future work in Section 5.

2.   MODELING LEAN AND DYNAMIC INVENTORY

2.1   Model of Lean Inventory

We model the lean inventory as consisting of n inventory locations. One item may be stored in each location. The inventory holds m different variants, with $m \leq n$, i.e. there are more inventory locations than there are variants. An inventory location may hold any one of these variants, and is not assigned to any particular variant. See figure 11-1. We assume that the probability of each occupied location holding each variant is $1/m$.

Such an inventory is denoted by *(n, m)*. We are interested in studying the customer service level of this inventory, which is defined as the average probability of a customer finding their desired variant in stock.

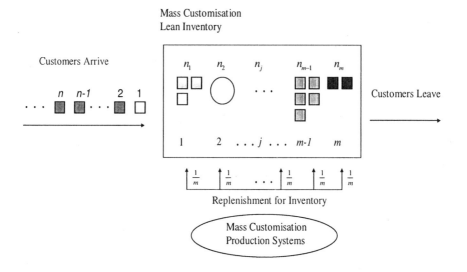

Figure 11-1. A Lean and Dynamic Inventory

The number of items of variant j, $1 \leq j \leq m$, that are stored in the inventory is denoted by n_j. For each variant, the number of items stored in inventory may range from zero to n, $0 \leq n_j \leq n$. However, the total amount of inventory must always be equal to n:

$$n_1 + n_2 + \ldots + n_m = n$$

Note that when a particular variant, j, is absent from inventory, $n_j = 0$. For example, in figure 11-1, n_2 is zero and is represented by a circle. In this case, instead of holding m different variants, the inventory holds only $m-1$ variants. Note that it is possible for more than one variant to be completely absent from inventory. However, the total number of items in stock must always equal n.

Current approaches to analyzing inventory focus on single item inventory. Many researchers propose Joint Replenishment Policy for multi-product inventory. Much of the literature pertaining to Joint Replenishment Policy only takes into account the optimization of the replenished item inventory in which the demand of each replenished item is dependent on the predetermined production cycle [6]. Until now, all models for single-product or multi-product inventories are based on cost and replenishment cycles, but do not take into consideration customers choosing from a number of variants and customer satisfaction.

Our literature reviews find that when customer service levels need to be appraised in the stage of designing or modeling inventory models, inventory designers or researcher simply assign them constants or percentages. This decision has two consequences: 1) to achieve the designed level, the cost is not optimal; 2) after finishing inventory designing, customer satisfaction is either higher or lower than expected. In industry, due to the absence of a simple model and tool to estimate customer service level, inventory managers rely on weekly or monthly stock checking to understand their performance in satisfying customers, which consumes company's resources and increases overhead costs for the inventory, especially in complex, multi-variant inventories.

We approach the problem from the perspective of combinatorial theory. This will enable us to obtain a mathematical expression for the probability of a customer finding their desired variant from any *(n, m)* inventory. In this way, the designer of an MC facility may balance the effects of the number of variants and the size of the inventory.

Our model allows the supply chain manager to determine an optimal inventory level with n units at any given time, given the number of product variants m on offer to provide a desired customer satisfaction of P_s.

2.2   Customer Arrival Pattern and Replenishment Policy

Customer arrival patterns and replenishment policies varied in many classic inventory models, such as the Economic Order Quantity model, Dynamic Lot Sizing model and Statistical Inventory model [4]. For this paper, we assume that a customer arrives and wishes to select one of m variants held in inventory. We assume that all variants are equally likely to be desired by the customer. If the customer's desired variant is available in inventory, it is withdrawn and immediately replaced by another item. In our paper, we set up the procedure of replenishment and explain how the items are selected for deliveries shown in figure 1. The replacement item is also selected randomly from the range of variants available, so that each replenishment item may also arrive with a probability of $1/m$. If the customer's desired variant is not available in inventory, the customer leaves without making a selection. We assume that only one customer arrives per replenishment cycle, so that an inventory of n items is always available to be searched for a customer's desired variant.

This inventory replenishment policy is similar to that found in, for example, car dealerships. Here, a limited stock of high value items is held, with some potential for late customization on site. The dealerships receive replenishment cars, but may not have complete control over which items are delivered to them. The arrival of customers is random, and is unlikely to follow the cycles described in this paper, but we shall address a random arrivals replenishment policy in a later paper. We discuss other extensions of this model, loosening the current assumptions, in the concluding section.

3.   MATHEMATICAL MODEL FOR CUSTOMER SERVICE LEVEL WITH LEAN INVENTORY

In this section, we will describe lean inventory in terms of the number of inventory locations, n, and the number of product variants, m. By applying combinatorial mathematics, the average probability of finding a customer's desired variant from the lean inventory can be calculated. An example of lean inventory with 10 inventory locations and three variants, i.e. a (10, 3) inventory, will illustrate the probability model.

3.1   Definition of terms

As mentioned in section 2.1, the number of each type of item that may be available in stock, n_j, may range from 0 to n. When any variant is not present in stock, i.e. $n_j = 0$, the customer's desired variant is unavailable and a sale is lost. If a customer arrives and all $n_j > 0$, then the customer will be able to find their desired variant. Once the customer has withdrawn their selection, a randomly selected variant replaces the withdrawn item. Since the selected item and replacing item are both randomly selected from the m variants, the number of at most two items in the stock will have changed, with one item increasing in quantity by one, and one item decreasing by one. Thus, if the customer selects variant j and the stock is replenished with item k, we expect:

$$_1n_j = {_0}n_j - 1;$$

$$_1n_k = {_0}n_k + 1 \qquad\qquad (1)$$

where the subscript to the left indicates the cycle number or timestep. Repeated applications of this withdrawal and replenishment policy will result in an inventory that evolves through various different states. When it happens that some of the variants are not represented in inventory, the average customer satisfaction level falls below 100%.

Here, we introduce the definition of i. i counts the number of variants that are absent from the current state of inventory. When all variants are present in inventory, $i = 0$. If one variant is absent, $i = 1$. So, i may take values from 0 to $m - 1$. When i takes the value $m - 1$, all locations in inventory are occupied by the same variant, l:

$$n_l = n$$

$$n_k = 0; \qquad k \neq l \qquad\qquad (2)$$

Our analysis of the customer satisfaction level proceeds by counting the number of possible ways in which n inventory items may be distributed over m variants, allowing for the possibility that there may be i variants that are absent from inventory. We define T_{total}, as the total number of possible ways of distributing these items over inventory:

$$T_{total} = f(n, m, i)$$ (3)

The level of customer satisfaction depends on the number of possible combinations that have $i > 0$, since this implies that one or more variants are absent from the inventory. We define T_i to be the number of combinations with i variants absent from inventory. Thus, T_0 is the total number of ways of distributing n items of inventory over the m variants such that all variants are represented, i.e. no variant is absent. Let T_1 be the number of ways of distributing inventory so that any one variant is missing from inventory. Similarly, we can list the values for T_2, T_3, ... , T_i, ..., T_{m-1}. Hence, T_{total} is given by:

$$T_{total} = T_0 + T_1 + \cdots + T_i \cdots + T_{m-2} + T_{m-1} = \sum_{i=0}^{m-1} T_i$$ (4)

This completes the definition of terms and explanation of the conceptual model. In the next section, we will develop the probability model.

3.2   Development of the Probability Model

In this section, we first consider the probability of finding a particular customer's desired variant. As discussed in Section 3.1, the inventory evolves into different distributions, which we shall refer to as states, according to the cycles of customers' arrivals and subsequent replenishments. When a particular customer arrives, he/she faces only one distribution or state of the inventory. The probability of successfully finding the desired variant is given by

$$P_{m,i} = \frac{m - i}{m}$$ (5)

Consider the (10,3) inventory that consists of 10 inventory locations holding some combinations of three variants. If two variants are absent from inventory, then $i = 2$, and the inventory contains one type of variant only. Hence, a customer who arrives at that inventory will have probability for finding their desired item of $1/3 = (3-2)/3$.

Having developed the expression for satisfying the customer's request, we must now consider the probabilities of occurrence of each of the T_i inventory distributions under the conditions described above of inventory replenishment. Since T_i is part of T_{total}, the probability of occurrence of T_i is given by the proportion of the total number of combinations, T_{total}:

$$P_{T,i} = \frac{T_i}{T_{total}} \tag{6}$$

For the lean, dynamic inventory, the average probability of finding a customer's desired variant is given by P_s, where P_s is the proportion of the inventory distributions with i absent variants multiplied by the probability of satisfying a random customer. Hence,

$$P_s = \sum_{i=0}^{m-1} P_{m,i} \times P_{T,i} \tag{7}$$

Substituting, we obtain the average probability for customers to find the desired product from the lean, dynamic inventory:

$$P_s = \sum_{i=0}^{m-1} \left(\frac{m-i}{m} \right) \times \left(\frac{T_i}{T_{total}} \right) \tag{8}$$

In the next section, we follow through the example of a lean, dynamic inventory with $n = 10$ and $m = 3$ to understand in more detail the meaning of $P_{T,i}$, $P_{m,i}$, P_s, T_{total} and T_i.

3.3   An Example of Lean Inventory (10,3)

In this section, we will develop the (10,3) example to show the occurrence of all the combinations of inventory. Refer to Table 11-1, which counts all possible ways of distributing the (10,3) inventory. Columns 2, 3 and 4 give the ways of distributing 10 items over three variants. Column 5 gives the multiplying factor that accounts for the different ways in which the items may be distributed over the three variants, denoted A, B and C. Thus, row 2 describes a combination that has 9 items of one variant, 1 item of a second variant and zero items for the third variant. This combination could be represented in inventory in six ways, as AAAAAAAAAB, AAAAAAAAAC, ABBBBBBBBB, BBBBBBBBBC, ACCCCCCCCC and BCCCCCCCCC. However, for notational convenience, we will denote this as (9,1,0), with $S_x = 6$ to allow for the distributions in separate variants. Note that the permutations of items within inventory are not relevant.

Table 11-1. Number of ways of distributing 10 inventory items over 3 variants

State x	Number of items of first variant n_1	Number of items of second variant n_2	Number of items of third variant n_3	Number of combinations S_x	Number of missing variants i
1	10	0	0	3	2
2	9	1	0	6	1
3	8	2	0	6	1
4	8	1	1	3	0
5	7	3	0	6	1
6	7	2	1	6	0
7	6	4	0	6	1
8	6	3	1	6	0
9	6	2	2	3	0
10	5	5	0	3	1
11	5	4	1	6	0
12	5	3	2	6	0
13	4	4	2	3	0
14	4	3	3	3	0
Total:				66	

Let us review how the lean inventory evolves. Customers arrive and find variants and the MC production system constantly delivers variants to replace those that the customer has withdrawn. The state of the inventory will constantly evolve through the states listed in Table 11-1. For example, suppose the inventory starts in state 1 or (10,0,0) where only one variant is currently held in stock. If a customer arrives and withdraws an item of that variant from stock, the inventory will temporarily be in state (9,0,0). However, the replenishment process immediately replaces the missing item, which may be the same as the other items in stock, or one of the other variants. Hence, the state of the inventory may now be either (10,0,0) or (9,1,0). But whatever the evolutionary stages, the state of the inventory will always be in one of the states listed in Table 11-1.

The last column in Table 11-1 shows the number of variants missing from each inventory state. For example, the number of missing variants, i, in states (10,0,0), (8,2,0) and (6,3,1) are 2, 1 and 0, respectively. Using the 14 states, we can calculate T_i. We can see from Table 11-1 that state x ($x = 1$)

has two missing variants. Hence, the number of combinations that have two missing variants is:

$$T_2 = 3$$

There are 5 states that have one missing variant, $i = 1$. These are $x = 2, 3, 5, 7, 10$. We sum up the corresponding S_x to obtain:

$$T_1 = 27$$

Similarly, we obtain for $i = 0$

$$T_0 = 36$$

We may now calculate the terms from Equation 7 and 8. For the case $i = 0$, we obtain:

$$P_{m,i} \times P_{T,i} = P_{3,0} \times P_{66,0}$$

$$= \left(\frac{m-i}{m} \right) \times \left(\frac{T_i}{T_{total}} \right)$$

$$= \left(\frac{3-0}{3} \right) \times \left(\frac{T_0}{T_{total}} \right) = 1 \times \frac{36}{66} \qquad (9)$$

$$= 0.5454$$

Similarly, if the inventory evolves into the distributions with 1 missing variant or 2 missing variants, similar analysis gives the probabilities:

$$P_{3,1} \times P_{T,1} = \left(\frac{m-1}{m} \right) \times \left(\frac{T_1}{T_{total}} \right) = \left(\frac{3-1}{3} \right) \times \left(\frac{27}{66} \right) = 0.2727 \qquad (10)$$

$$P_{3,2} \times P_{T,2} = \left(\frac{m-2}{m} \right) \times \left(\frac{T_2}{T_{total}} \right) = \left(\frac{3-2}{3} \right) \times \left(\frac{3}{66} \right) = 0.0152 \qquad (11)$$

Summing up the probabilities from Equations 9, 10 and 11 and we obtain:

$$P_s = \sum_{i=0}^{m-1} \left(\frac{m-i}{m}\right) \times \left(\frac{T_i}{T_{total}}\right) = 0.8333 \tag{12}$$

Thus, in lean inventory (10, 3), the average probability of finding the customer's desired variant is 0.8333 and we may say that the customer service level is 83.33%.

Through the above analysis on lean inventory example (10,3), we obtain experience of the methodology of how to obtain the values of $P_{T,i}$, $P_{m,i}$, P_s, T_{total} and T_i while $n = 10$ and $m = 3$. In figure 11-2, we summarize the steps in calculating T_{total} and T_i for inventory (10,3) and develop a procedure for obtaining these two values for any lean inventories with n inventory items and m kinds of variants $(n \geq m)$. See figure 11-2.

Begin

Step 1 **Enumerate** all states of distributions of inventory (n, m) exhaustively and completely.

Step 2 **Compute** combinations S_x for each state of inventory.

Step 3 **Sort and categorize** all states according to the number of missing variants, i, in the inventory. The states with same number i of missing variants should be categorized into one group ($0 \leq i \leq m-1$).

Step 4 **Sum up** combinations S_x of each state for each group with i missing variants to obtain T_i. Then, $T_{total} = \sum_{i=0}^{m-1} T_i$.

End

Figure 11-2. Summary of Procedure to Calculate T_{total} and T_i

This procedure will help us obtain the value of T_{total} and T_i for any inventory (n, m). Then, the probability of finding a customer's desired variant from lean inventory can be calculated by Equation 8. The advantage of this procedure is that it enables us to calculate the combinations for every particular state of the inventory. However, it is not straightforward to compute the number of states T_i when i may range from 0 to $m-1$. We re-evaluate the situation in the next section and propose a new solution for T_{total} and T_i from a different perspective.

3.4 Extension of Liu's Combinatorial Mathematics to Complete the Final Expression for Customer Service Level

Liu's combinatorial mathematics simplifies our problem to "n indistinct balls are assigned to m distinct boxes" [5]. It is very relevant to our research problem and may be applied to find the exact values of T_{total} and T_i in the inventory with any values of n and m. We review the work in his book and extend it to complete our final expression for the probability of finding a customer's desired variant from lean MC inventory.

As discussed in Section 2.1, n units can be visualized as including n_1 units of one variant, n_2 units of another variant, ..., and n_j units of the jth variant (where $j \leq m$). For each variant, all items are the same so permutations do not need to be considered.

Since we can first distribute one unit in each of m boxes (we call cells) and then distribute the remaining $n - m$ units arbitrarily, the number of ways of distributing is

$$C((n-m)+m-1, n-m) = C(n-1, n-m) = c(n-1, m-1) \qquad (13)$$

This result can be extended to calculate the number of ways of distributing n non-distinct units into m distinct cells with each cell containing at least q units. In Equation 13, $q = 1$. After placing q units in each of the m cells, we will distribute the remaining $n - mq$ units arbitrarily, and then the number of ways of distributing is

$$C((n-mq)+m-1, n-mq) = C(m-mq+n-1, m-1) \qquad (14)$$

For the example inventory (10, 3), we substitute $n = 10$, $m = 3$ and $q = 0$ into Equation 14:

$$C(m-mq+n-1, m-1) = C(12, 2) = 66$$

This confirms the result in Table 11-1. We have also validated this for other inventory cases. Equation 14 is applicable for all positive integers, n and m, in the case of $n \geq m$.

The total number of ways of distributing, or the number of states of inventory, T_{total}, can be derived from Equation 14 with $q = 0$:

$$T_{total} = C(m+n-1, m-1) \tag{15}$$

Furthermore, we consider empty cells in the inventory, equivalent to missing variants. We introduce the parameter i into Equation 14 to calculate T_i. When there are no missing variants in the inventory, $i = 0$, meaning that each cell has at least 1 variant and $q = 1$, therefore

$$T_0 = C(m-mq+n-1, m-1) = C(n-1, m-1) \tag{16}$$

For example, in lean inventory (10, 3):

$$T_0 = C(n-1, m-1) = C(9, 2) = 36$$

which is same as the value from our procedure in Section 3.3.

If there is one empty cell in the inventory, $i = 1$, this is equivalent to n units distributed into $m - 1$ cells where each cell of $m - 1$ cells (excluding the empty cell) contains at least $q = 1$ item. The number of ways of distributing derived from Equation 14 is

$$C(m-1-(m-1)q+n-1, m-1-1) = C(n-1, m-2) \tag{17}$$

Since each cell in the inventory can be that empty cell, the total number of ways to be the empty cell is $C(m, 1)$. Therefore,

$$T_1 = C(m, 1) \times C(n-1, m-2) \tag{18}$$

Similarly, we can extend this calculation for $i = 2, 3, \ldots, m - 1$, the number of ways to have i empty cells is $C(m, i)$. The number of ways of distributing n units into $m - i$ cells and $m - i$ cells contain at least q items, is:

$$C(m-i-(m-i)q+n-1, m-i-1)$$
$$= C((m-i)(1-q)+n-1, m-i-1) \tag{19}$$

Therefore

$$T_i = C(m, i) \times C((m-i)(1-q)+n-1, m-i-1) \tag{20}$$

Equation 4 can be written as

$$T_{total} = \sum_{i=0}^{m-1} T_i = \sum_{i=0}^{m-1} C(m,i) \times C((m-i)(1-q)+n-1, m-i-1) \qquad (21)$$

Except for the empty cells, all other cells must have at least $q = 1$ item in the cell. So, Equation 20 and 21 will be

$$T_i = C(m,i) \times C(n-1, m-i-1) \qquad (22)$$

$$T_{total} = \sum_{i=0}^{m-1} T_i = \sum_{i=0}^{m-1} C(m,i) \times C(n-1, m-i-1) = C(m+n-1, m-1) \,(23)$$

Substituting Equations 22 and 23 into Equation 8, we can generate the expression for the average probability of finding a customer's desired variant from MC lean inventory:

$$P_s = \sum_{i=0}^{m-1} \left[\left(\frac{m-i}{m} \right) \times \left(\frac{T_i}{T_{total}} \right) \right]$$

$$= \frac{1}{C(m+n-1, m-1)} \sum_{i=0}^{m-1} \left[\left(\frac{m-i}{m} \right) \times C(m,i) \times C(n-1, m-i-1) \right]$$

$$(n \geq m) \qquad (24)$$

Equation 24 is our final mathematical expression for customer service level in MC lean and dynamic inventory shown in figure 11-1.

In mass customization organizations, the resources allocated to inventory are limited. However, the level of inventory must maintain a high number of product variants in order to satisfy customers' diverse demands. The customer service level will be influenced by inventory locations n and the number of variants m. Inventory managers can make decisions for the maintenance of inventory based on the relationship of n, m and P_s through Equation 24. More details follow in the next section.

4. RELATIONSHIP BETWEEN CUSTOMER SERVICE LEVEL, INVENTORY CAPACITY AND PRODUCT VARIETY

In this section, we analyze the relationship between customer service level, inventory capacity and product variety using Equation 24. We have defined the probability of finding the desired variant, P_s, as the customer service level of MC lean and dynamic inventory. Inventory units n and m kinds of variants represent inventory capacity and product variety. First, we give fixed values of m and substitute them into the Equation 24 to obtain Figure 11-3. Second, fixed values of n are used to obtain Figure 11-4. These two graphs describe the relationship of P_s, n and m. We also discuss the implications for the inventory manager to maintain lean and dynamic MC inventory.

As shown in Figure 11-3, we give fixed numbers of variants, 3, 10, 20 and 30 and the corresponding probabilities are calculated by Equation 24. The horizontal axis represents inventory units n and the vertical axis represents the average probability of finding the desired variant. While $m = 3$, and remembering that $n \geq m$, n are given from 3 to 100. The distribution of probabilities ranges from 60% to 98%. We can see an obvious curve when n takes small values and the distribution flattens while n increases. The results shows inventory managers exactly how many more units should be kept to successfully satisfy customers when the number of variants increases. For example, if lean inventory aims to achieve customer service level at 80%, the inventory should have approximately 34, 75 and 116 inventory units, when m takes values 10, 20 and 30, respectively. In the cases of $n = m$, customer service level of inventory (3, 3) is around 60%. However, in inventory (10, 10), (20, 20) and (30, 30), probabilities are around 50%. We estimate that customer service levels are around 50% for any big values of n and m, where $n = m$.

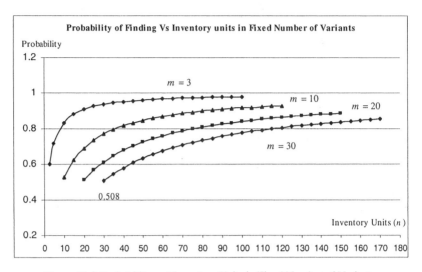

Figure 11-3. Probability and Inventory Units in Fixed Number of Variants

Figure 11-4 shows the distributions of probabilities over different number of variants when there are fixed values of inventory units. For inventory units, 10, 20 and 30, the probabilities decrease when numbers of variants in the inventory increase. The probabilities of small *n* decrease much quicker than for large *n*. For any values of *n*, the probabilities of finding a customer's desired variant is 1 when there is only one kind of variant in the inventory, i.e. *m* = 1.

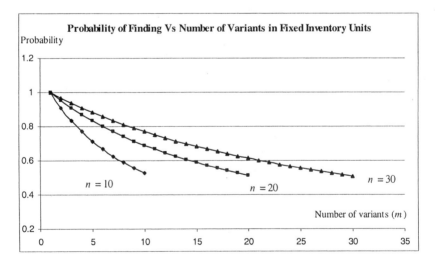

Figure 11-4. Probability and Number of Variants in Fixed Inventory Units

In this section, we obtain probability distributions at different n and m through the analysis of Equation 24. Based on the analysis, inventory managers can make decisions about maintaining lean and dynamic MC inventory. If they want to have a high customer service level, they can arrange bigger inventory spaces to increase the capacity when the number of variants keeps unchanged. Accordingly, more holding cost and maintenance cost will be incurred for this MC lean inventory. Another way to have a high customer service level is to have a reasonable product variety instead of some large m. This action may limit customers' choices when they come into the inventory and lost sales may result. When the inventory is at the customer involvement point, the industrial engineer may decide to locate the customer involvement point where the number of variants is sufficiently low for a lean inventory to satisfy the customer, while not taking up too much space or cost. Another efficient way to achieve high customer service level is for inventory managers to maintain a responsive lean and dynamic inventory. If they observe some variants are not that popular, they can stop such variants' delivery and send some popular variants to increase the profitability and customer service level. Our mathematical model and analysis give a very clear relationship of customer service level, inventory capacity and product variety.

5. SUMMARY AND CONCLUSION

It is a reality for manufacturing enterprises to maintain semi-finished and finished goods inventories. In future, mass customization strategies may become industry standard through the effort in academia and industry. Ideally, there would be no stock in MC organizations. However, in next 20 or 30 years, the issue of inventory cannot be avoided and inventories are still necessary for the majority of manufacturing organizations. What we can do is make such inventories lean and cost-effective in the context of mass customization.

In this paper, we model a lean and dynamic inventory and develop a mathematical expression to understand the relationship between customer service level, inventory capacity and product variety. Mass Customization organizations will not spend too much budget and resources on the inventories. The capacity or space of inventory is limited. High product variety is a key successful driver for any MC organizations. It is main difference between mass customization and mass production. Our mathematical model will help understand the balance between inventory capacity and product variety, in the lean and dynamic inventory with a special customer arrival pattern and replenishment policy.

We successfully apply combinatorial mathematics in developing a mathematical expression for the customer service level of lean inventory. This enables inventory practitioners to understand the impact on the customer service level from limited inventory capacity and high product variety in mass customization organizations. For example, if managers want to achieve targeted customer service level, they can balance inventory capacity and product variety. They can either increase inventory capacity and have more inventory units, or decrease product variety and have fewer kinds of variants in the inventory. They can tailor economical inventories with proper inventory capacity and reasonable product variety themselves on the basis of customers' demand and customer service level.

Our future work will give new customer arrival patterns. Accordingly, the delivery from MC production systems to lean inventory will be more responsive to customers' demand. In each cycle, more than one customer may arrive. The delivery may have time delay and takes the policy of Lot Sizing replenishment. We will conduct case studies in car dealership and supermarkets to compare with our theoretical model. We will build up simulation model for lean and dynamic MC inventory and measure its performances, such as complexity and cost.

REFERENCES

Davis, S. (1989): From future perfect: Mass Customizing, Planning Review 17(2) (1989) 16 – 21.

de Alwis A. / Mchunu C. / Efstathiou J. (2002): Methodology for Testing Mass Customisation Strategies by Simulation *Proceedings of the International Conference of the Production and Operations Management Society (POMS - San Francisco 2002)*. 14th-8th April 2002.

Hopp, W. J. / Spearman, M. L. (2001): Factory Physics, Foundations of Manufacturing Management.

Lampel, J. / Mintzberg, H. (1996): Customizing Customization, in: Sloan Management Review/Fall (1996) 21-29.

Liu, C. L. (1968): Introduction to Combinatorial Mathematics, Department of Electrical Engineering, Massachusetts Institute of Technology, McGraw-Hill, Inc.

Siajadi, H. / Ibrahim, R. N. / Lochert, P. B. / Chan, W. M. (2005): Joint replenishment policy in inventory-production systems, *Production Planning & Control*, Apr2005, Vol. 16 Issue 3, p255, 8p.

Chapter 12

HRM POLICIES FOR MASS CUSTOMIZATION

Understanding individual competence requirements and training needs for the mass customizing industrial company

Cipriano Forza[1], Fabrizio Salvador[2]
[1] *Universitá di Padova;* [2] *Instituto de Empresa Business School*

Abstract: Management research has been long emphasizing the fundamental contribution of individual competencies to the development of company competences, which thus affect organizational success or failure. To date, the debate over critical individual competences has been almost ignored in research on Mass Customization. As a consequence, we know a lot about the possible methods and organizational arrangements supporting Mass Customization, but we know very little as to how the human variable enters the puzzle. Given this fact, the present paper explores what roles are mostly affected by mass customization within the company, as well as what fundamental requirements Mass Customization poses to the manufacturing firm in terms of individual competences. We address these questions by means of a qualitative interview-based study involving 46 experts across 5 European countries. Our findings indicate that (1) R&D operative roles play a central role in the achievement of a Mass Customization capability and suffer from a serious gap or customization-related competencies; (2) individual attitudes, almost neglected in extant research on Mass Customization, are essential in determining the individual's capability to perform customization-related tasks and finally (3) that methods knowledge is perceived by respondents as not so critical as individual attitudes and context knowledge.

Key words: Mass Customization, Competences, Human Resources, Knowledge, Training

1.   INTRODUCTION

Undeniably, product and service customization is becoming more and more a competitive must, rather than a curious oddity (Alstrom and West-brook, 1999; Pine, 1993). Such trend is so pervasive that we now take it for granted we can personalize a car, a vacation, or an insurance package, according to our own needs. Despite the multi-disciplinarily and variety of research on mass customization (see Blecker et al., 2005; Duray, 2002; Tseng and Piller, 2003; Salvador et al., 2004), past research has been almost exclusively focusing on identifying and formalizing *methods* to efficiently customize products and services (e.g. see Forza and Salvador, 2002; Huang et al., 2005; Jiao and Tseng, 2000). Methods are here intended as techniques, process structures, organizational arrangements and technologies supporting some aspect of mass customization, meant to be valid for multiple companies. Turning methods into a "mass customization capability" however, takes more than merely transferring methodological knowledge from books to people. At the very least, we should understand what kind of company-specific knowledge individuals need to embed into their working practices. In addition, it may be that certain behavioral traits of human personality, such as friendliness or open-mindedness, are more critical than others to the successful deployment of mass customization methods. Last but not least, we do not know whether mass customization-related knowledge and behavioral attitudes differ across sales, operations and R&D. As a matter of fact, there is a total lack of research specifically addressing these topics, which are only tangentially mentioned in a few works (e.g. see Bramham and MacCarthy, 2005; Hirschhorn et al., 2001; Kakati, 2002).

The present paper aims at exploring the behavioral components, individual knowledge and individual abilities needed to support a company-wide mass customization capability. This goal has been pursued by means of a qualitative research involving interviews to 46 managers experts on the topic across different five European countries. In the next sections we report the framework underlying data collection and analysis, then describe the sample of companies and informants, we detail the qualitative data analysis procedures followed, and we discuss the messages emerging from interview data. We finally speculate on the broad implications of the study findings as well the opportunities it opens to future research.

2.   RESEARCH FRAMEWORK AND QUESTIONS

Research on individual competences provides the conceptual foundations to explore the relation between individuals' characteristics and the firm mass customization capability. This body of knowledge, in fact, conceptualizes what are the inner constituents of the individual's competences, as well as how and why these constituents can contribute to organizational effectiveness. A major obstacle inherent any borrowing from the literature on individual's competences is the lack of a unified theory, as a myriad of alternative perspectives and definitions have been proposed. Different definitions of competency can be valid, to the extent that such differences are motivated by specific characteristics of the context where the definition is used (Hoffman, 1999). Taking Hoffmann's (1999) suggestion, we selectively picked from the literature a set of concepts helpful to structure our research as well as to give meaning to the picture we get from our empirical inquiry. These concepts are: *individual competence, individual ability, individual knowledge* and *individual attitude.*

Necessarily, our starting point is the *(individual) competence*, which we define as a set of individual characteristics determining the individual's capability to successfully perform a given job within a specific work environment. This definition is based on the notion of competence proposed by Tett et al. (2000) and is coherent with reference work by academics (see MacClelland, 1973), practitioners (see Boyatzis, 1982), and by the European Centre for the Development of Vocational Training, the official European Union's reference centre for vocational education and training.

Even though every competence is made of specific individual characteristics, all competences can be thought as made of three broad classes of individual characteristics, namely *abilities, knowledge* and *attitudes.*

We define *ability* as the capability to perform a certain physical or mental task. This definition was originally proposed by Spencer and Spencer (1993) to designate what is a "skill". Skills and abilities, however, are mostly used as interchangeable terms within the industrial and organizational psychology literature, to the extent that concepts labeled as 'skills' in certain taxonomies are labeled as 'abilities' in other works. Compared to competences, abilities describe something more specific and detailed, as they *focus on a specific task.* In terms of their constituents, abilities can be seen as *deriving from a bundling of attitudes and knowledge.* For example, the ability of planning may be though as resulting from knowledge of planning principles and an attitude towards order, quality and accuracy. Consequently, *abilities can be developed,* mainly by acquiring specific *knowledge* and, if possible, by fostering or correcting certain *attitudes.*

The third central concept is that one of *knowledge.* We refer to knowledge as what an individual has to know to be able to perform a task (see Mirabile, 1997; Nonaka, 1994).

Finally, we define *attitude* as a tendency of an individual to act in a consistent way to a particular object or situation, as proposed by Fishbein and

Ajzen (1975). Attitudes, therefore, are about the behavioral characteristics of an individual, i.e. how a person acts, thinks and feels (Ackerman, 1988).

Based on the literature, therefore, the relation between individual competences, attitudes, abilities and knowledge can be formalized as portrayed in Figure 12-1.

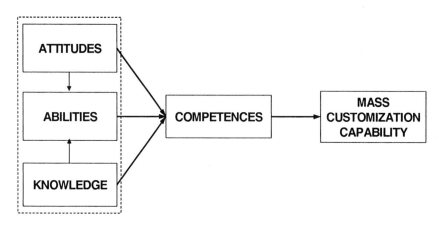

Figure 12-1. Research framework

This theoretically-derived competence framework allows us to explore how mass customization affects managerial and operational roles within the various functional areas of the company. More specifically we address the following questions: (1) which professional roles' competencies are mainly affected by customization? (2) What individual abilities, attitudes and knowledge are specifically needed for Mass customization? And, finally (3) what are the required training needs?

3.   METHOD

Given the absence of any published research on the topic, we decided to adopt a qualitative approach to explore the individual competences supporting the organization capability for mass customization. This is in line with the basic goals of qualitative research as it allowed us to explore an unknown territory without excluding aprioristically non-evident messages and areas of inquiry (Miles and Huberman, 1994). Hereafter we discuss our sampling approach, interview protocol and process as well as data analysis procedures.

3.1   Sample

In selecting the companies we tried to maximize heterogeneity while retaining the possibility to make meaningful comparisons. As for heterogeneity is concerned, we included companies from five different European countries: Germany, Italy, Slovenia, Spain and United Kingdom. Second, they differ in size, both in terms of number of employees and net sales. Third, we tried to avoid developing insights specific to certain kinds of companies by including in the sample companies making product with a very wide price spectrum, and therefore products with different degree of complexity. In addition, products included in the sample have different production volume and, therefore, production processes with different degree of repetitiveness.

As for sample homogeneity is concerned, we restricted the sample to enterprises below 1,000 employees, because, due to higher task specialization, larger enterprises are likely to place different competence requirements upon their personnel. Secondly, we constrained the sample to manufacturing companies. The resulting sample is shown in table 12-1.

In terms of respondents, we tried to have a multiplicity of perspectives, especially from three fundamental areas involved in the delivery of customized goods: sales and marketing, operations (production and logistics) and research and development, which in customization-oriented companies is mostly focused on development. In the case of small companies, the key "customization expert" is the owner or general director, so that in these cases we just interviewed this key informant.

3.2   Interview process and protocol

The exploratory nature of our research prompted us to rely on a quasi-unstructured questionnaire for data collection. The questionnaire was based on a few open-ended main questions, essentially mirroring the research questions illustrated in Section 3.

Table 12-1. Company sample - part A

Number of employees	Percentage	Country	Percentage	Net sales (millions €)	Percentage
< 50	18%	IT	29%	<1	11%
50 - 99	25%	ES	29%	1 - 9,9	25%
100 - 244	25%	UK	14%	10 - 49	29%
250 – 1.000	32%	GER	14%	50 - 150	32%
		SL	14%	>150	4%
N.A.	0%	N.A.	0%	N.A.	0%
TOTAL	100%	TOTAL	100%	TOTAL	100%

Table 12-2. Company sample - part B

Turnover from person-alized products	Percentage	Pieces per person (annual)	Percentage	Euro per piece	Percentage
0% - 10%	14%	0 - 9	11%	< 1	4%
11% - 50%	29%	10 - 99	21%	1 - 9,9	18%
51% - 90%	29%	100 - 999	14%	10 - 99	18%
91% - 100%	18%	1.000 - 9.999	18%	100 - 999	11%
		10.000 - 99.999	11%	1.000 - 9.999	25%
		100.000 - 857.143	7%	10.000 - 46.603	7%
N.A.	11%	N.A.	18%	N.A.	18%
TOTAL	100%	TOTAL	100%	TOTAL	100%

Table 12-3. Respondents sample

Respondent role/position	Number	Percentage
General director/entrepreneur	11	24%
Marketing/sales director	8	17%
Technical director	10	22%
Operations (production, logistic) director	12	26%
Human resource manager	2	4%
Others	3	7%
TOTAL	46	100%

In order to make sure all essential data from the interviews was collected, an "interview report form" and a "respondent characteristics form" were prepared.

3.3 Data analysis procedure

The early part of qualitative data analysis was performed by all project members, and consisted in cutting from the interviews significant excerpts that either highlighted key roles and competencies for mass customization or related specific competencies to specific roles.

Identifying the key roles sets affected by mass customization required some interpretation. Every respondent, in fact, used company-specific names to designate organizational positions within the various areas of the firm. We firstly clustered the mentioned roles, (e.g. sales manager, post sales manager, area sales manager, marketing manager, product manager, communication manager), into "roles sets."

Synthesizing from interviews excerpts the competencies mostly affected by customization was a second complex step of qualitative data analysis. Even tough we tried to stick to already available taxonomies this was not actually possible. The reason for the impossibility to fully adopt past taxonomies is essentially due to their generality. Being general, they do not include sufficient detail as for customization specific knowledge, abilities and methods. We derived from interview excerpts ad-hoc categories (see Rubin and Rubin, 1995; Kvale, 1996) in order to capture such specific knowledge, abilities and methods.

4. ROLES AFFECTED BY MASS CUSTOMIZATION

The first fundamental question we explored is whether customization evenly affects all the key roles sets considered in our study or if, instead, some roles are more heavily affected than the others. In other words, we tried to understand to what extent certain roles require the acquisition and mastering of more customization-specific competencies than others.

In table 12-3 the professional figures (column 3) cited as affected by customization are grouped in role sets (column 1). To quantify how much the various role sets are thought to be affected by product customization column 4 reports the percentages of interviews that cite the role set as influenced by product customization while column 5 shows the level of influence i.e. the level with which each role set is affected by product customization. The level of influence is measured on a scale 0 to 3 where 0 = "not influenced at all" (and therefore is not cited as a roles set affected by customization), 1 = "a little bit influenced", 2="moderately influenced", 3="highly influenced".

Table 12-3 highlights that, in managers' and entrepreneurs' opinion, product personalization influences the various role sets differently, depending on the function and on the hierarchical level. More specifically, the role sets influenced by product customization are those which belong to the operational functions, namely Research & Development, Marketing & Sales, Production and Logistic functions.

Marketing/sales function is that one that globally is more influenced by product personalization. This is due to the fact that this area has big responsibilities in defining product assortment and in communicating effectively product offer to the customer during order acquisition. At the same time, a wide, difficult-to-acquire spectrum of competences is needed by Sales & Marketing people to fully contribute to the company's mass customization capability.

R&D operative roles, i.e. product engineers and designers, overall, are extremely influenced by product personalization and are influenced at the same level when compared with marketing /sales role sets. The criticality of R&D operatives has a twofold interpretation. On the one hand, it underscores the pivotal part played by these people in determining a company-wide organizational mass customization competence. Due to the multiple implications of product design for marketing, logistics, production and R&D decisions taken by designers, i.e. R&D operatives, may have serious and wide implications for the company capability for mass customization. On the other hand, interviewers appeared to consider R&D operatives so critical because the traditionally narrow focus of engineers' training and evaluation tends to make their competence gap particularly severe. The R&D operatives' competence gap is further exacerbated by behavioral shortcomings of the typical R&D operative profile. They often appear to be rigid, narrow-focused and unresponsive to commercial priorities and needs.

Operative roles of production and logistics are though to be minimally affected by product customization. This can be expected for Logistics operational roles: the most adverse situation is that they are involved in some customized packaging activity. In the case of Production operational roles, instead, this message appears to go against some of the common wisdom embedded in the literature on non-repetitive manufacturing. We would expect, in fact, customization to disrupt efficiency in any manufacturing operation, slowing down learning curves and increasing the risk of operator errors due to task variability (Schonberger, 1986; Pentland, 2003). As multiple production directors made clear, however, most of the burden of customization can be relieved from Production operatives by appropriate product industrialization and disciplined production planning (see Dean & Snell, 1991).

*Table 12-4.*Professional figures and role sets affected by product customization

Role Set (Acronym)	Professional figures	Percentage of citations in interviews	Level of influence
Marketing & Sales Directors (MSD)	Sales manager, Sales administrator, Post sales manager, Area manager, Communications Manager, Marketing manager, Marketing, Technical marketing manager, Product Manager	67%	1,93
Marketing & Sales Operatives (MSO)	Post sales assistant, Assistant sales manager, Sales, Commercial representative, Sales representative, salesman, Sales Engineer – Field support engineer, Product configurator helpdesk, Applications Engineer, Production estimator, Customer Service, Customer liaison manager	59%	2,02
Technical Directors (TD)	Head of technical office, R&D manager, R&D project manager, Project manager, Project team manager, Designer team leader, Special project coordinator, Senior special project manager, Special product design manager, Technical coordinator	35%	1,22
Technical Operatives (TO)	Senior product design engineer, R&D design technician, R&D technician, Project engineer, Special project engineer, Project team engineer, Special product design engineer, Process design engineer, Industrialization technician, Manufacturing engineer, Manualist	63%	1,93
Production Directors (PD)	Head of production, Production Manager, Manufacturing Manager, Production supervisor, Supervisor, Head of tool workshop	59%	1,74
Production Operatives (PO)	Production technician, Maintenance technician, Machine set-up technicians.	9%	0,22
Logistic Directors (LD)	Logistics, Logistics Manager, Production Planning manager, Head buyer, Purchasing, Distribution manager	35%	1,07
Logistic Operatives (LO)	Logistic operator	2%	0,04
Other (O)	IT system engineer, CRM development role, Human resources manager, Finance manager, General director	11%	0,30

5.   MASS CUSTOMIZATION COMPETENCES

The following step in the analysis of the interviews is performed to identify the competences mostly affected by product customization i.e those which received a higher number of citations. The percentages reported in Figures 12-2, 12-3 and 12-4 indicate the ratios between the total number of citations a competence received divided by the total number of citations received by the class of competence (attitudes, abilities, knowledge) to which the competence belongs.

First of all we can notice that 29% of citations fall in the class of attitudes while 38% belong to the class of knowledge and 33% to the class of abilities. The interviewees therefore consider that customization affects all the components of competences, namely attitudes, knowledge and abilities. Looking at the differences between the three classes we can notice a greater incidence of the knowledge which is quoted 9% times more than attitudes. We can therefore argue that interviewees consider customization as putting a little bit more requirements in terms of knowledge than in terms of attitudes.

In order to gain a deeper understanding on how product personalization affect individual competences it is necessary to analyze the detail of each class of competence. For this purpose we executed a Pareto analysis within each class of competence.

The attitudes most influenced by customization are those related to the need to face a more complex, uncertain and ever changing environment. In particular we notice that Mind Flexibility (30%), Negotiation and relationships orientation (17%), Team working attitudes (cooperate and promote cooperation and sense of team) (14%), Creativity and innovative thinking (11%) are those that are considered most affected by customization (i.e. together reach 70% of total citations within their class).

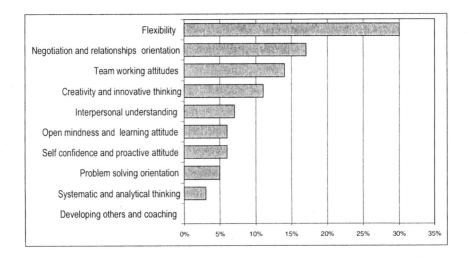

Figure 12-2. Attitude categories affected by mass customization

Moving to Knowledge we can notice that methodological knowledge is far less considered that contest knowledge. In fact the interviewees quoted most often Knowledge of product from technical point of view (14%), Knowledge of product applications (13%), Knowledge derived by breath of experience (10%), Knowledge of production processes (10%), Knowledge of company capabilities (8%), Knowledge of Technical/production issues (7%), and Knowledge of Methods to evaluate cost and financial implications (7%).

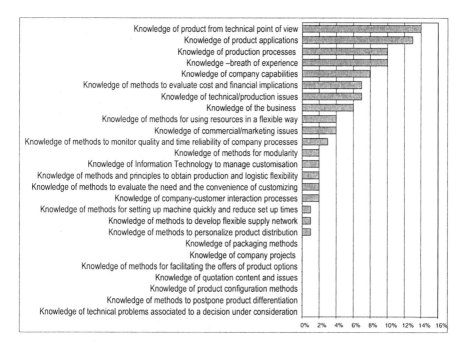

Figure 12-3. Knowledge categories affected by mass customization

Finally by considering the abilities we can notice that the interviewees consider as most affected those abilities that are directly related to the order specification, acquisition and fulfillment process. Able to think and act in a customer-oriented way (14%), Able to act considering technical/production issues associated to each decision (8%); Able to act evaluating cost and financial implications of each decision (7%); Able to understand rapidly and correctly each specific customer (7%); Able to obtain collaboration (6%); Able to plan, coordinate and organize (6%); Able of using resources in a flexible way (5%); Able to act considering commercial/marketing issues (5%); Able to anticipate technical problems (5%); Able to evaluate the need and the convenience of customizing (4%); Able to use all the available information before initiating an action (4%) are considered the most influenced abilities. The relatively high number of abilities mentioned underline that the employee when working in a customization context puts several different action requirements in the everyday work thus requiring a different profile of several individual abilities.

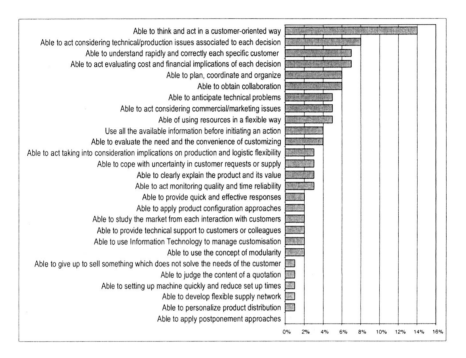

Figure 12-4. Ability categories affected by mass customization

6.   TRAINING NEEDS TO BUILD MASS CUSTOMIZATION

Once identified the competences which are thought to be affected by product customization we moved to consider what competence training managers and entrepreneurs consider necessary to augment the success in product personalization. Figures 12-5, 12-6 and 12-7 highlight these training needs. Column percentages indicate how often each training need has been mentioned, compared to the total number of citations received by the class of training needs to which it belongs. A high value for a given training need indicates that a) the relative competence is thought to be influenced by product customization, b) that managers and entrepreneurs envisage that this competence is less present than it would be desirable for mass customization, and

c) the competence is though to be improvable through training. Point C pinpoints the key difference between mass customization training needs and mass customization critical competences, as described in Section 5.

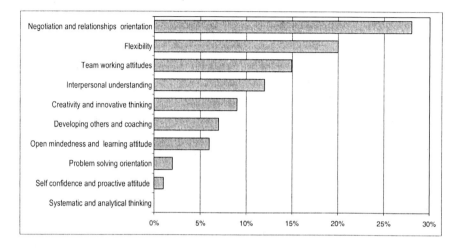

Figure 12-5. Attitudes that have to be developed

First of all we can notice that 23% of citations of competences training needs fall in the class of attitudes while 44% belong to the class of knowledge and 33% stay in the class of abilities. This time the three class of competences have been quoted much differently thus indicating either a different level of gap or a different possibility to act through training to reduce the gap.

Firstly the fact that Knowledge training needs are counted twice than attitudes training needs may indicate that managers and entrepreneurs, while considering both attitudes and knowledge very important (see table 12-3), consider Knowledge more suitable to be transferred through training than attitudes. Attitudes may be obtained in an easier way through appropriate people selection even though the 23% of quotation indicates that some positive results may be obtained through training too.

Secondly the fact that abilities are very frequently quoted suggests that managers and entrepreneurs perceive a need to develop abilities that can immediately affect customization performance. In other terms, they implicitly highlight the need for a training in those competences that have an non-mediated effect on performance.

By looking at the perceived need for attitudinal training, it emerges that interviewees see training needs mostly to foster efficacy of interpersonal relationships. In fact improving Negotiation and relationships orientation

(28%), Mind Flexibility (20%), Team working attitudes (15%) and Interpersonal understanding (12%) support one individual's capacity to better relate to other persons.

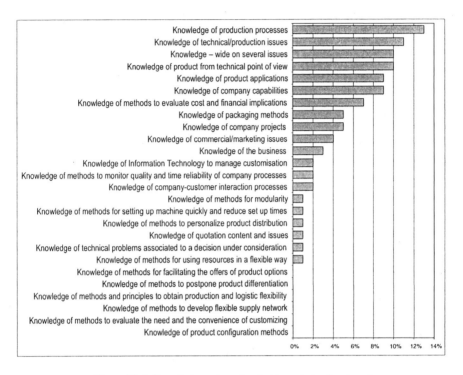

Figure 12-6. Knowledge categories that have to be developed

A more careful look at knowledge reveals as crucial training on contextual knowledge. Methods knowledge, on the other side, is not perceived as highly affected by customization neither and coherently there is limited perception of training need on this aspect of knowledge. Training needs, therefore, emerge for Knowledge of production processes (13%), Knowledge of technical/production issues (11%), Knowledge of product from technical point of view (10%), Knowledge – wide on several issues (10%), Knowledge of company capabilities (9%), Knowledge of product applications (9%). The only method knowledge which emerges with Pareto analysis is Knowledge of methods to evaluate cost and financial implications (7%).

By looking at abilities it emerges that managers and entrepreneurs highlighted a training need for the abilities required to an efficient and effective order definition, acquisition and fulfillment process. It emerges the need to train employees for a multidimensional set of abilities: mainly a combination of commercial and technical abilities with the addition of some financial and organizational abilities. Interviewees, in fact, highlighted the training need for the following abilities: Able to act considering technical/production issues associated to each decision (15%), Able to act evaluating cost and financial implications of each decision (9%), Able to understand rapidly and correctly each specific customer (8%), Able to think and act in a customer-oriented way (8%), Able to plan, coordinate and organize (8%), Use all the available information before initiating an action (7%), Able to act considering commercial/marketing issues (5%), Able to study the market from each interaction with customers (5%), Able to provide quick and effective responses to customers and colleagues (4%), and Able to clearly explain the product and its value (4%).

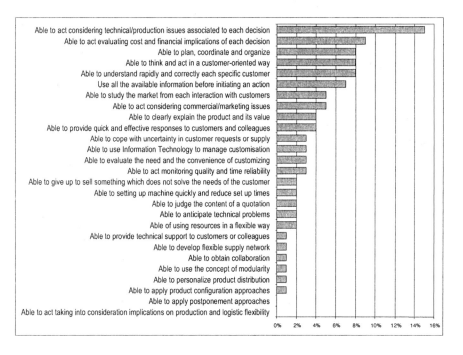

Figure 12-7. Abilities that have to be developed

7. THE TRAINING NEEDS FOR EACH ROLE SET

Different roles sets appear to have different training needs either in term of the total amount of training required and in terms of the specific profile of training needs. We concentrated on the first aspect to identify which role sets present a higher need for training. One way to derive this information is to consider the sum on all interviews of the number of competences that each interviewee considers to necessitate improvement through training for a given role set. Table 12-5 reports both the absolute number of citations and the percentage calculated on the total number of citations.

Table 12-5. The role set citation in terms of competence training needs

Role set	Acronym	Citations	Percentage
Marketing & Sales Directors	MSD	92	17%
Marketing & Sales Operatives	MSO	126	24%
Technical Directors	TD	58	11%
Technical Operatives	TO	116	22%
Production Directors	PD	90	17%
Production Operatives	PO	5	1%
Logistic Directors	LD	39	7%
Logistic Operatives	LO	0	0%
Other	O	9	2%
TOTAL		535	100%

Table 12-6 highlights that Marketing & Sales operative roles, i.e. salesmen, (24%) and technical operative roles, i.e. product engineers and designers, (22%) are the roles most in need of training. Marketing/sales directive roles (17%) and production directive roles (17%) need less training and technical directive roles (11%) even less. The other professional roles show almost no training need with the exception of the logistic directive roles (7%) that show very limited needs. Globally the commercial function represents 41% of the total training needs while technical function represents 33% of total training needs and the production-logistic directive roles covers almost all the remaining needs with a 24% of the total training needs.

In order to increase the confidence on these results we added to these figures the analyses of the training process. We calculated the number of interviewees that cited each role set when specifying which professional figures should be included in the training initiatives. Table 12-7 shows the results of this analysis.

Table 12-6. The role set citation in terms of participation of training processes

Role set	Acronym	Number of interviews citing the role set	Percentage (100% = 46 interviews)
Marketing & Sales Directors	MSD	20	43%
Marketing & Sales Operatives	MSO	19	41%
Technical Directors	TD	11	24%
Technical Operatives	TO	23	50%
Production Directors	PD	22	48%
Production Operatives	PO	6	13%
Logistic Directors	LD	10	22%
Logistic Operatives	LO	1	2%
Other	O	4	9%

Technical operative roles (50%) are again the main cited ones but this time are followed by production directive roles (48%) and at less extent by marketing/sales directive (43%) and operative (41%) roles. Similarly to the previous analyses technical directive roles are positioned at a lower level (24%) as the logistic directive roles (22%) even though they are not negligible. Production operatives roles (13%) receives some more consideration than in the previous analysis but still remain at low levels while logistic operative roles remain negligible. This analysis therefore leads to the same indications than the previous one even though it emphasizes the training needs of the production function.

These results constitute a further evidence that marketing/sales operatives and technical operatives are the two crucial roles for product customization, since they not only are the most affected by customization but also are those that mostly need to learn and improve their competences.

8.   CONCLUSIONS, LIMITATIONS AND FURTHER RESEARCH

We believe our exploratory inquiry on the role individual competencies for mass customization brings to the attention of the research community a topic that has been left outside mainstream research on mass customization. Our study revealed a number of messages. First, our inquiry indicates operational roles in Research & Development, together with Marketing and Sales roles are the crucial ones to reach a mass customization capability. Second, our study identifies a set of specific attitudes, knowledge and abilities that combined together build the customization capability an individual possess. Surprisingly, methods knowledge seem to play a secondary role in comparison with attitudes. Third, the roles that are mostly affected by customization are the same roles that are identified as mostly in need of training. The analysis of training needs highlighted the fundamental importance of the production director. When comparing training needs with competence needs, the importance of training programs aiming at knowledge transfer is emphasized compared to the importance of attitudinal training. This is in line with the widely accepted fact that it takes more to modify attitudes than to transfer knowledge. It is interesting that the attitudes most underlined as needed to be trained regards the interaction between individuals. It is also notable that the knowledge considered affected by customization and the knowledge seen as that one to be transferred in training is the contextual knowledge. All the reported findings bear the limitations of qualitative research. Accordingly, these finings will be subsequently subjected to empirical validation in the next phase of the study, which is survey-based.

REFERENCES

Ackerman, P. L., 1988, Determinants of individual differences during skill acquisition: Cognitive abilities and information processing, *Journal of Experimental Psychology: General*, **117**(3):288-318.

Alström, P., and Westbrook. R., 1999, Implications of mass customization for operations management - An exploratory survey, *International Journal of Operations and Production Management*, **19**(3):262-274.

Blecker, T., Abdelkafi, N., Kaluza, B., Friedrich, G., 2005, Controlling variety-induced complexity in mass customization – a key metrics based approach, *International Journal of Mass Customization*, forthcoming.

Boyatzis, R. E., 1982, *The Competent Manager. A model for Effective Performance*, John Wiley & Sons, New York.

Bramham, J., MacCarthy, B., and Guinery, J. 2005, Managing product variety in quotation processes, *Journal of Manufacturing Technology and Management*, **16**(4):411-431.

Dean, W. J., and Snell, A. S., 1991, Integrated manufacturing and job design: moderating effects of organizational inertia, *Academy of Management Journal*, **34**(4):778-804.

Duray, R., 2002, Mass customization origins: mass or custom manufacturing?, *International Journal of Operations and Production Management*, **22**(3):314-328.

Fishbein, M., and Ajzen, I., 1975, *Belief Attitude Intention and Behavior: An introduction to Theory and Research*, Addison-Wesley, Reading.

Forza, C., and Salvador, F., 2002, Product configuration and inter-firm coordination: an innovative solution from a small manufacturing enterprise, *Computers In Industry*, **49**(1):37-46.

Hirschhorn, L., Noble, P., and Rankin, T., 2001, Sociotechnical systems in an age of mass customization, *Journal of Engineering and Technology Management*, **18**(3-4):241-252.

Hoffmann, T., 1999, The meanings of competency, *Journal of European Industrial Training*, **23**(6):275-285.

Huang, G. Q., Zhang, X. Y., and Liang, L., 2005, Towards integrated optimal congfiguration of platform products, manufacturing processes, and supply chains, *Journal of Operations Management*, **23**(3-4):267-290.

Jiao, J., and Tseng, M. M., 2000, Understanding product family for mass customization by developing communality indices, *Journal of Engineeering Design*, **11**(3):225-243.

Kakati, M., 2002, Mass customization - needs to go beyond technology, *Human Systems Management*, **21**(2):85-93.

Kvale, S., 1996, *Interviews - An Introduction to Qualitative Research Interviewing*, Sage Publications, Thousand Oaks.

McClelland, D. C., 1973, Testing for competence rather than for "intelligence", *American Psychologist*, **28**(1):1-14.

Miles, M. B., and Huberman, M. A., 1994, *Qualitative Data Analysis: An Expanded Sourcebook*, Sage Publications, Thousand Oaks.

Mirabile, R. J., 1997, Everything you wanted to know about competency modeling, *Training & Development*, **51**(8):73-77.

Nonaka, I., 1994, A dynamic theory of organizational knowledge creation, *Organization Science*, **5**(1):14-37.

Pentland, B., 2003, Conceptualizing and measuring variety in the execution of organizational work processes, *Management Science*, **49**(7): 857-870.

Pine, J. B. II., 1993, *Mass Customization - The New Frontier in Business Competition*, Harvard Business School Press, Cambridge.

Rubin, H. J., and Rubin, I. S. 1995, *Qualitative Interviewing - The Art of Hearing Data*, Sage Publications, Thousand Oaks.

Salvador, F., Rungtusanatham, M., and Forza, C., 2004, Supply chain configurations for Mass Customization, *Production Planning & Control*, **15**(4):381-397.

Shonberger, R. J., 1986, *World Class Manufacturing*, Free Press, New York.

Spencer, L. M., and Spencer, S. M., 1993, *Competence at Work. Models for Superior Performance*, John Wiley & Sons, New York.

Tett, R. P., Guterman, H. A., Bleier, A., and Murphy, P. J., 2000, Development of a content validation of a "hyperdimensional" taxonomy of managerial competence, *Human Performance*, **13**(3):205-251.

Tseng, M. M., and Piller, F. T., 2003, *The Customer Centric Enterprise: Advances in Mass Customization and Personalization*, Springer Verlag, Berlin.

INDEX

CONTRIBUTORS

Abdelkafi, Nizar

is a PhD candidate and research fellow at the Hamburg University of Technology, Department for Business Logistics and General Management (5-11), Schwarzenbergstr. 95, 21073 Hamburg, Germany. He worked within the interdisciplinary multi-year research projects "Modeling, Planning, and Assessment of Business Transformation Processes in the Area of Mass Customization" and "TECTRANS -Technology Transfer, both University of Klagenfurt. He holds an industrial engineering diploma from the National Engineering School of Tunis, Tunisia, and a Master in Business Administration from the Technische Universität München, Germany. Nizar Abdelkafi is co-author of the book "Information and Management Systems for Product Customization". email: nizar.abdelkafi@tu-harburg.de, homepage: http://web.logu.tu-harburg.de

Arokiam, Ivan

Phd, Msc(Eng), B. Eng.) Agility Centre Project Manager: The University Of Liverpool. Ivan's background is in Mechanical, Manufacturing Systems and Lasers. He has worked in companies including Golden Penny Flour Mill, The Coca-Cola Company Ltd and The UK Centre for Materials Education. Ivan's current role in the Agility Centre entails him working with many small companies in a drive to improve manufacturing performance. Some of the publications including those in which he co-authors include areas of Laser cutting, applications of cellular manufacturing and mass customisation. His interests include design for manufacture and operations. Agility Centre, University of Liverpool Management School (ULMS), University of Liverpool, PO Box 147 Liverpool L69 7ZH

Blecker, Thorsten

is full professor at the Hamburg University of Technology, Department for Business Logistics and General Management (5-11), Schwarzenbergstr. 95, 21073 Hamburg, Germany. He holds a masters degree in business administration (with honors) and a PhD (summa cum laude) from the University of Duisburg, Germany. He finished his habilitation thesis in September 2004 at the University of Klagenfurt, Austria. Thorsten Blecker is guest-editor of a special issue of IEEE Transactions on Engineering Management on "Mass Customization Manufacturing Systems" (forthcoming), co-editor and author of several books, e.g. "Production/Operations Management in Virtual Organizations", "Enterprise without Boundaries", "Competitive Strategies", "Web-based Manufacturing" and "Information and Management Systems for Product Customization". Main research interests: business logistics and supply chain management, production/operations management, Industrial information systems, internet-based production systems, mass customization manufacturing systems, strategic management, and virtual organizations. Homepage: http://www.manufacturing.de/, http://web.logu.tu-harburg.de, email: blecker@ieee.org.

Bock, Stefan

Lecturer at the Graduate School, University of Paderborn, Faculty of Economics, War-burger Straße 100, 33098 Paderborn, Germany, Email: stbo@upb.de, Biographical notes: 1970: Born in Bielefeld (Germany), 1991-1996: Studies of Computer Science and Business Administration at the University of Paderborn, 1996: Graduated as "Diplom Informatiker", 1997: Received Faculty Award for the best Diploma Thesis of the year 1996 at the Faculty of Computer Science, 1999: PhD in Production Management. Title of PhD Thesis: "Models and distributed algorithms for planning assembly lines", 2000: Received Industrial PhD Thesis Award of the Unternehmensgruppe Ostwestfalen e.V., 2003: Habilitation Degree in Business Administration (German Post Doc Degree). Title of Professorial dissertation: "Real-time control of freight forwarder transportation networks", Since 2004: Lecturer for Business Computing in the Graduate School Dynamic Intelligent Systems of the University of Paderborn.

Brabazon, Philip

is a Research Fellow in the Operations Management Division on the Nottingham University Business School. His area of research is the Mass Customization of products and services and his interests include the development of quantitative and qualitative operational templates. Email: philip.brabazon@nottingham.ac.uk.

Culley, Steve

is Reader and Head of Design in the Department of Mechanical Engineering at Bath University. He has researched in the engineering design field for many years. In particular this has included the provision of information and knowledge to support engineering designers. He is a Fellow of the Institution of Mechanical Engineers and Non Executive Director of Adiuri Systems. Department of Mechanical Engineering, University of Bath, Claverton Down, Bath BA2 7AY, UK.

Efstathiou, Janet

is a Reader in Engineering Science at the University of Oxford, and Tutorial Fellow in Mechanical Engineering at Pembroke College. Dr Efstathiou established the Manu–facturing Systems Group to carry out research in the area of manufacturing complexity, the supply chain and mass customisation. She has over one hundred publications in control, manufacturing and complexity. Janet Efstathiou obtained a BA in Physics with History and Philosophy of Science at University of Oxford and a PhD in Computing from the University of Durham. She has held lectureships in the University of London (Department of Electrical and Electronic Engineering) and in the University of Cambridge (Computer Laboratory). Department of Engineering Science, University of Oxford, Parks Road, OXFORD OX1 3PJ, U.K, web: http://www.robots.ox.ac.uk/~manufsys

Forza, Cipriano

holds M.Sc. degree in Electronic Engineering from Padova University (Padova, Italy) and PhD degree in Industrial Management from Padova University. He has been visiting scholar at Minnesota University, London Business School and Arizona State University. He is Professor of Management and Operations Management at Modena e Reggio Emilia University and Padova University. He has been called to hold the position of full professor in Management and Engineering at Padova University, starting from November 2005. He is faculty of EDEN seminars for OM PhD's of European Institute of Advanced Studies in Management and he is faculty of the PhD course in Management and Engineering of Padova University. He currently serves as regional secretary of the Italian Association of Management and Engineering. He is responsible of the Mass Customization and Information System Sections of the International Research Project on High Performance Manufacturing. Currently he serves as associate editor of Journal of Operations Management and is member of the Editorial Board of Decision Science. He researched on management of logistical flows within and across companies, quality management, Information Systems supporting operations, operational performance and operations improvement, and research methods in Operations Management. Currently his research focus is on mass customization and product variety management. Università degli Studi di Modena e Reggio Emilia, Dipartimento di Ingegneria Meccanica e Civile, Via Vignolese, 9005/a, - 91100 Modena, - Italy, E-mail: forza.cipriano@unimo.it

Friedrich, Gerhard

is professor at the Department of Computer Science and Manufacturing, University of Klagenfurt, Austria. He holds a masters degree in computer science and a PhD from the Technical University of Vienna, Austria, where he also finished his habilitation in 1994. He was a visiting scientist at the Stanford Research Institute and the corporate research center of Siemens. Gerhard Friedrich worked for several years in the industry as the head of the department for Object-oriented and Knowledge-based Configuration and Diagnosis Systems, Siemens AG, Austria. He initiated and contributed to several international research projects, e.g., "CAWICOMS – Customer-Adaptive Web Interface for the Con–figuration of Products and Services with Multiple Suppliers". He is guest-editor of a special issue of IEEE Transactions on Engineering Management on "Mass Customization Manufacturing Systems" (forthcoming) and co-author of the book "Information and Management Systems for Product Customization". Gerhard Friedrich is member of the advisory board of the "International Journal of Mass Customization", member of the editorial board of AI Communications, and member of the board of advisors of Configworks, a software company in the field of personalized handling and servicing of customers via various distribution channels. Main research interests: personalization of web-based information systems, configuration systems, mass customization, diagnosis, knowledge-based systems. email: gerhard.friedrich@ifit.uni-klu.ac.at, Homepage http://www.ifi.uni-klu.ac.at/IWAS/GF/.

Hotz, Lothar

is a researcher at the Hamburger Informatik Technologie Center (HITeC) located at the University of Hamburg. He participated in several projects related to topics of knowledge-based configuration, knowledge representation, constraints, diagnosis, qualitative simulation, parallel processing and object-oriented programming languages. HITeC e.V., Universität Hamburg, Vogt-Kölln-Str. 30, 22527 Hamburg, Germany, email: hotz@informatik. uni-hamburg.de.

Huang, George Q.

received the BEng and Ph.D. degrees in mechanical engineering from Southeast University in China and Cardiff University in the UK in 1983 and 1991, respectively. Dr. Huang is an Associate Professor at The University of Hong Kong. His main research interests include platform products for mass customization, supply chain configuration, grid design and manufacturing, and computational game theory. He has published extensively in these topics, including over 70 journal papers, two monographs entitled Cooperating Expert Systems in Mechanical Design and Internet Applications in Product Design and Manufacturing respectively, and an edited reference book entitled Design for X: Con–current Engineering Imperatives. Dr. Huang is a Chartered Engineer, and a member of IEE (UK), HKIE, IIE and ASME. Department of Industrial and Manufacturing Systems En–gineering, University of Hong Kong, Hong Kong, P. R. China, E-mail: gqhuang@hku.hk.

Ismail, Hossam

PhD BSc, MIEE, CEng, Senior lecturer at The University of Liverpool Management School. Hossam has numerous publications in the areas of Manufacturing Systems Design, Simulation and Modelling, Agility, Mass customisation, Decision Support Tools, Intelligent Design and Feature recognition. Hossam is Director of the Agility Centre, a business help centre funded from Objective One for Merseyside to develop agility tools and methodologies to assist manufacturing-based Merseyside-based SME's. Hossam has supervised eight knowledge transfer programmes and he has four major Research and Development projects in design and manufacturing. Hossam is also responsible for a successful MSc programme in e-business strategy and systems at The Management School; a programme that is specifically aimed at manufacturing industry. Agility Centre, University of Liverpool Management School (ULMS), University of Liverpool, PO Box 147 Liverpool L69 7ZH.

Jannach, Dietmar

is University Assistant at the Institute of Business Informatics and Application Systems, University of Klagenfurt. University of Klagenfurt, Institute of Business Informatics and Application Systems, Computer Science and Manufacturing, Universitätsstrasse 65-67, 9020 Klagenfurt, Austria.

Jørgensen, Kaj

is associate professor at Dept. of Production, Aalborg University. His primary research area is Information Modeling applied to Product Configuration and Product Modeling and Building Modeling. At his department, he is the coordinator of the research group Information Technology in Production Systems and he is currently member of the International Advisory Committee of the annual international conference Engineering Design & Automation (EDA). His primary teaching subjects are Information Modeling, Information Systems Development, Product Modeling, Product Configuration and Product Meta Data Modeling. Web.: www.iprod.auc.dk/~kaj, Email: kaj@iprod.aau.dk.

Krebs, Thorsten

is a researcher at the Laboratory for Artificial Intelligence (LKI) at the University of Hamburg. He has participated in developing the configuration tool EngCon at the Centre for Computing Technologies (TZI) at the University of Bremen. Key interests are model-based configuration and knowledge representation. Current work addresses (dynamic) evolvability of knowledge. Universität Hamburg, Vogt-Kölln-Str. 30, 22527 Hamburg, Germany, email: krebs@informatik.uni-hamburg.de.

Kreutler, Gerold

is a PhD candidate and research assistant at the Department of Computer Science and Manufacturing at the University of Klagenfurt, Austria within the interdisciplinary multi-year research projects "Modeling, Planning, and Assessment of Business Transformation Processes in the Area of Mass Customization" and "TECTRANS – Technology Transfer". He holds a master degree in computer science (with honors) and is working in the domain of configuration systems for several years, especially in consideration of online customer advisory. He has taken part in several consultancy projects for the implementation of Enterprise Resource Planning systems. He is co-author of the book "Information and Management Systems for Product Customization". Main research interests: personalization of web-based information systems, business process management and the application of ERP systems. email: gerold@kreutler.net, Homepage: http://www.kreutler.net/gerold/.

Lehne, Ernesto Del Valle

is currently in the process of finalising his research at Wolfson College, University of Oxford and working as a Logistics Planner for a first tier service provider at BMW's MINI plant in Oxford. He obtained his BEng in Industrial and Systems Engineering in 1995. From 1998 he has worked in Logistics and Supply Chain, in areas ranging from sea freight to OTR transport at Maersk Sealand. He came to the UK to do an MSc in Systems Control at Sheffield University in 2000. Department of Engineering Science, University of Oxford, Parks Road, OXFORD OX1 3PJ, U.K, web: http://www.robots.ox.ac.uk/~manufsys.

Lu, Wuyi

is a third year D.Phil student in Pembroke College, University of Oxford. He obtained his B.Eng in Electromechanical Engineering in 1999. His research interests include Mass Customisation, Manufacturing Complexity, Supply Chain and Lean Inventory. From Oct 2002 to Nov 2003, he worked as a research assistant in the Department of Engineering Science, University of Oxford. He used to be a business consultant, specialised in transportation & logistics in McKinsey & Consulting. He was a supply chain production leader in a sportswear trading and retailing company and a production engineer in a casting & processing company. Department of Engineering Science, University of Oxford, Parks.

MacCarthy, Bart

is Professor of Operations Management at Nottingham University Business School and Director of the Mass Customization Research Centre (MCRC). As well as Mass Customisation, his research interests include the analysis and design of operational systems in business and industry with particular emphasis on responsiveness and time compression across the extended enterprise. He has researched and consulted with a wide range of industries including textiles and clothing, automotive, engineering, aerospace, consumer products and food, as well as with firms in distribution and logistics. He is a Fellow of IEE, The Institute of Mathematics and its Applications and the Institute of Operations Management. He has published widely on Operations Management, Management Science and related areas. Email: bart.maccarthy@nottingham.ac.uk.

McIntosh, Richard

received his PhD on "The Impact of Innovative Design on Fast Tool Change Methodologies" from University of Bath in 1998. Since then he is working as a Research Officer at the University of Bath on changeover improvement. Currently he is also undertaking research into mass customisation. Department of Mechanical Engineering, University of Bath, Claverton Down, Bath BA2 7AY, UK

Mileham, A R

is Head of the Department of Mechanical Engineering at University of Bath and undertakes research in the general area of Manufacturing with particular reference to the modelling and optimisation of manufacturing processes and systems. His manufacturing systems research has been in the areas of assembly systems, computer aided process planning, cost estimation and rapid changeovers. Department of Mechanical Engineering, University of Bath, Claverton Down, Bath BA2 7AY, UK

Owen, Geraint

received his Phd on the "Design of Transfer Lines" from University of Bath. Subsequently employed as a Research Officer, still at Bath, working mainly on Rapid Changeover, but also spending time doing a survey for the design council and also the design of an experimental rig for the Russian MIR space station. Increased involvement in the teaching activities led to an appointment as a Teaching Fellow in 1996, and subsequently a Lecturer in 2000. Department of Mechanical Engineering, University of Bath, Claverton Down, Bath BA2 7AY, UK

Poolton, Jenny

PhD MBA BSc(Psychol) MBPS MCIM. Jenny acts as Marketing Manager and Marketing Analyst within The Agility Centre based at The University of Liverpool Management School. Jenny graduated with a BSc degree in Psychology from the University of Warwick and a PhD in New Product Development from the University of Liverpool. She also has an MBA in International Business and Finance from Oklahoma University. Since graduation, Jenny has worked as an academic and has published widely. Her first love though is working practically with companies. Since joining the Agility Centre, Jenny has been actively working with over 35 SME's seeking to improve their new product development and marketing performance. Agility Centre, University of Liverpool Management School (ULMS), University of Liverpool, PO Box 147 Liverpool L69 7ZH

Reid, Iain. R.

(M.Eng,B.Eng (Hons)). Iain Reid is the Senior Project Manager within the Agility Centre, University of Liverpool. His role is the promotion and implementation of 'best practice' of 'Lean' and 'Agile' manufacturing techniques within SME's. His first degree is from the Sheffield Hallam University in B.Eng (Hons) Design Manufacture with Management and with an M.Eng in Manufacturing Systems Engineering. His research interests and publications include Mass Customization, Business Process Redesign, Knowledge Management, New Product Development (NPD), Customer-Driven manufacturing such as Engineer-to-Order. He is currently developing a framework for ETO Project-Based learning as part of his PhD. Agility Centre, University of Liverpool Management School (ULMS), University of Liverpool, PO Box 147 Liverpool L69 7ZH.

Reik, Michael

studied engineering at University of Karlsruhe (TH), Germany and received his MSc in Advanced Mechanical Engineering from Imperial College London in 2003. Since early 2004 he is a researcher at University of Bath and writing his Phd on Design for Changeover. Department of Mechanical Engineering, University of Bath, Claverton Down, Bath BA2 7AY, UK, M.Reik@bath.ac.uk

Salvador, Fabrizio

is Professor of Operations Management at Instituto de Empresa (Madrid, Spain) and Research Associate at Arizona State University (Tempe, Arizona). He holds an MS in Engineering and Management and a PhD in Operations Management from Università di Padova (Padova, Italy). His research interests are related to mass customization, product configuration and the relation between product design and organization design in environments characterized by high product variety. He regularly teaches these topics to undergraduates, postgraduates and executives. He has been successfully assisting nume–rous companies in managing product variety-related problems. His work has been publi–shed in prestigious academic and practitioner journals, including the Journal of Operations Management, International Journal of Operations and Productions Management, Computers in Industry, International Journal of Production Economics, Business Horizons and other journals. He recently published a book with McGraw-Hill on product con–figuration. Instituto de Empresa, Department of Operations and Technology Management, Maria de Molina, 12 - 28006 Madrid – Spain, E-mail: Fabrizio.Salvador@ie.edu

Wolter, Katharina

is a research assistant at the Laboratory for Artificial Intelligence at the Department of Computer Science, University of Hamburg. She works in the EU project "Configuration in Industrial Product Families" (ConIPF). Her research interests are in the area of knowledge-based configuration and human-computer interaction, especially exploratory configuration. Her current work includes undo support for interactive configuration and user-centered product configuration. Universität Hamburg, Vogt-Kölln-Str. 30, 22527 Hamburg, Germany, email: kwolter@informatik.uni-hamburg.de

Zhang, Xin Yan

received her B.E. and M.S. degrees from Huazhong University of Science and Technology, Wuhan, China, in 1999 and 2002, respectively. Miss Zhang is now studying her PhD degree in the Department of Industrial and Manufacturing Systems Engineering, the University of Hong Kong. Her research interests include platform products for mass customization, supply chain configuration, computational game theory, computer aided process planning, and product data management.

Early Titles in the
INTERNATIONAL SERIES IN
OPERATIONS RESEARCH & MANAGEMENT SCIENCE
Frederick S. Hillier, Series Editor, *Stanford University*